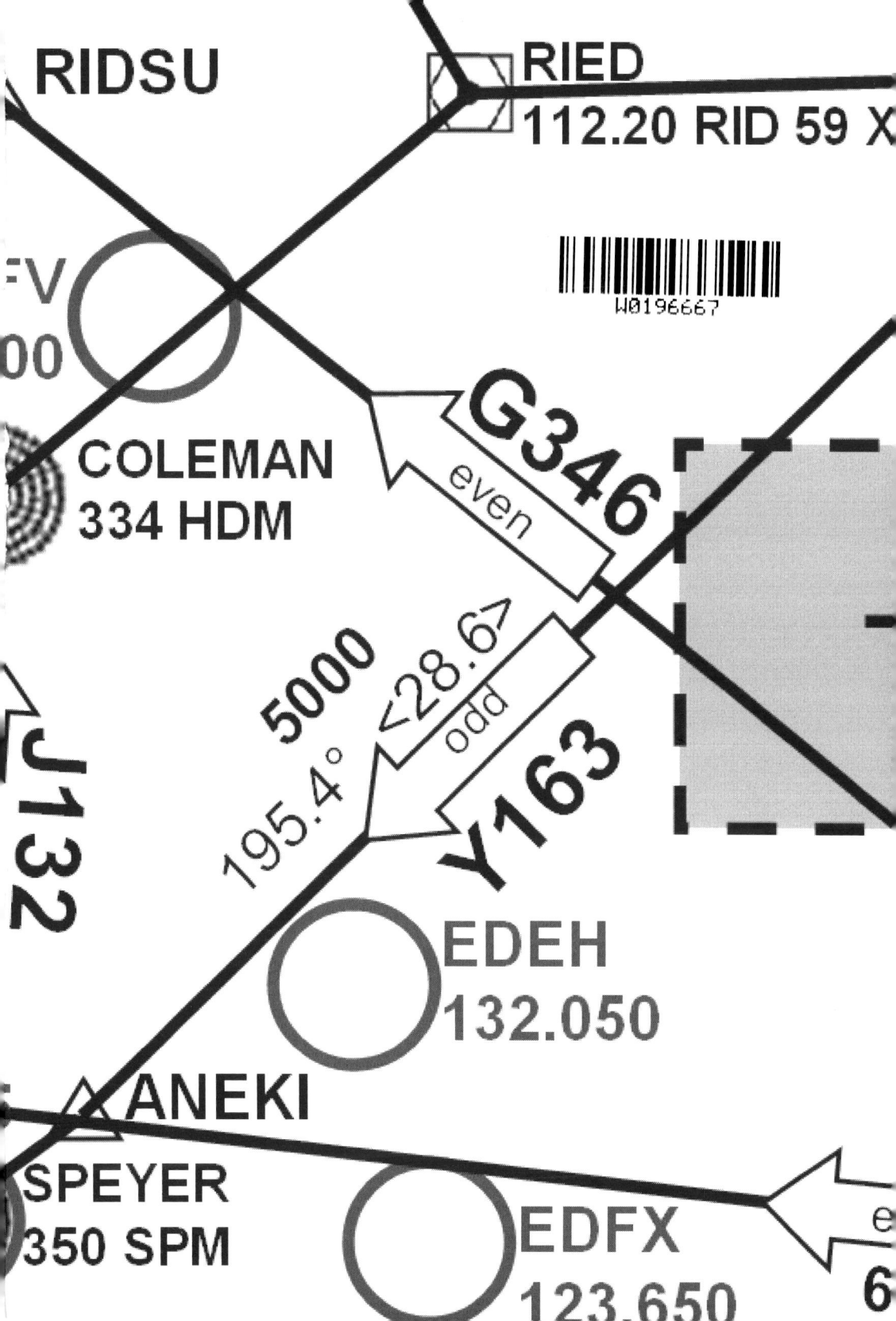

aeromedConsult Hinkelbein Neuhaus GbR

Goethestr. 7

76771 Hördt

http://www.aeromedconsult.de

eMail: info@aeromedconsult.de

Jochen Hinkelbein • Christopher Neuhaus

Prüfungsvorbereitung für die Privatpilotenlizenz

Band 8B:

Allgemein gültiges Sprechfunkzeugnis (AZF)

2. Auflage 2009

Impressum

Zitierweise

Hinkelbein, Jochen / Neuhaus, Christopher: Prüfungsvorbereitung für die Privatpilotenlizenz, Band 8B: Allgemein gültiges Sprechfunkzeugnis (AZF), 2. Auflage, Rev. 2, November 2009, *aeromedConsult* Hinkelbein Neuhaus GbR, ISBN-13: 978-3-941375-08-6

2. Auflage 2009: ISBN-13: 978-3-941375-08-6

1. Auflage 2008: ISBN-13: 978-3-0002-4847-4

© *aeromedConsult* Hinkelbein Neuhaus GbR, Goethestr. 7, 76771 Hördt

Umschlaggestaltung:

Christopher Neuhaus

Coverfoto (Buchumschlag):

Dr. med. Jochen Hinkelbein

Vertrieb und Bestelladresse:

aeromedConsult Hinkelbein Neuhaus GbR, Goethestr. 7, 76771 Hördt

Bestellfax: +49 (0) 3212 1184690

Web: http://www.aeromedconsult.de

eMail: info@aeromedconsult.de

Vorwort

Eine gute Sprechfunkausbildung ist die Basis für eine erfolgreiche Luftfahrertätigkeit. Dies gilt uneingeschränkt sowohl bei Privatpiloten für das beschränkt gültige Sprechfunkzeugnis (BZF), als auch für Berufs-flugzeugführer oder Verkehrspiloten, die täglich regen Sprechfunkverkehr betreiben müssen und bei denen ein Allgemeines Sprechfunkzeugnis (AZF) erforderlich ist.

Das vorliegende Buch mit dem Titel "**Band 8B: Allgemein gültiges Sprechfunkzeugnis (AZF)**" aus der Reihe "Prüfungsvorbereitung für die Privatpilotenlizenz" in der nunmehr **2. Auflage** soll als Hilfe zur Prüfungsvorbereitung und zur effektiven Wiederholung des relevanten Stoffs für die Theorieprüfung dienen.

In der überarbeiteten **2. Auflage** ist es an den offiziellen **Fragenkatalog des Jahres 2009** angelehnt und erklärt alle Fragen und Antworten ausführlich und verständlich. Zusätzlich erleichtern vielen Grafiken bzw. Tabellen den Lernerfolg. Das Ziel beim Lernen ist daher nicht nur zu wissen, welche Antwort richtig oder falsch ist, sondern auch zu verstehen, warum das so ist.

Für weitere Anregungen sind wir stets dankbar! Nur so kann eine kontinuierliche Weiterentwicklung des Buches gewährleistet werden.

Wir wünschen allen Lesern viel Erfolg bei der Prüfungsvorbereitung und insbesondere bei der theoretischen Prüfung zum AZF!

Mannheim und Heidelberg im November 2009

Dr. Jochen Hinkelbein und Christopher Neuhaus

Bearbeitungshinweise

Das vorliegende Buch soll als Hilfe zur Prüfungsvorbereitung und zur effektiven Wiederholung des relevanten Stoffs für die Theorieprüfung dienen. Hierzu ist es eng an den offiziellen **Fragenkatalog 2009** angelehnt und erklärt alle Fragen und Antworten ausführlich und verständlich.

Auf der linken Buchseite ist jeweils die Original-Prüfungsaufgabe aus dem PPL-Fragenkatalog samt Antwortmöglichkeiten (1) wieder gegeben. Die Nummerierung der Fragen (2) stimmt mit der des Fragenkatalogs überein. Buchstaben in Klammern nach der Frage (3) beziehen sich auf die relevanten Prüfungsaufgaben bzw. den relevanten Lernstoff für die jeweilige Lizenzart:

- A = PPL-A (JAR-FCL)
- C = PPL-C (Segelflugzeugführer)
- DG = PPL-D (Freiballone Gas)
- DH = PPL-D (Freiballone Heißluft)
- E = PPL-H (Hubschrauber)
- N = PPL-N (national)
- KS = Kontrollierter Sichtflug (CVFR)

Geänderte oder neue Fragen werden im Buch separat mit einem grauen Balken neben der Prüfungsfrage kenntlich gemacht.

Auf der rechten Buchseite sind jeweils die kommentierten Antworten zu den Fragen aufgelistet (4). Unterschiedliche Symbole (5) helfen bei der Differenzierung zwischen „richtig" und „falsch".

Viele Fragen und Sachverhalte sind zusätzlich durch didaktisch aufbereitete Bilder bzw. Tabellen (6) verdeutlicht, um das Verständnis zu erleichtern

 Besonders wichtige Sachverhalte bzw. Kernaussagen werden am Ende der Seite separat kenntlich gemacht und zum besseren Verständnis nochmals gesondert und kurz <u>auf deutsch</u> erklärt.

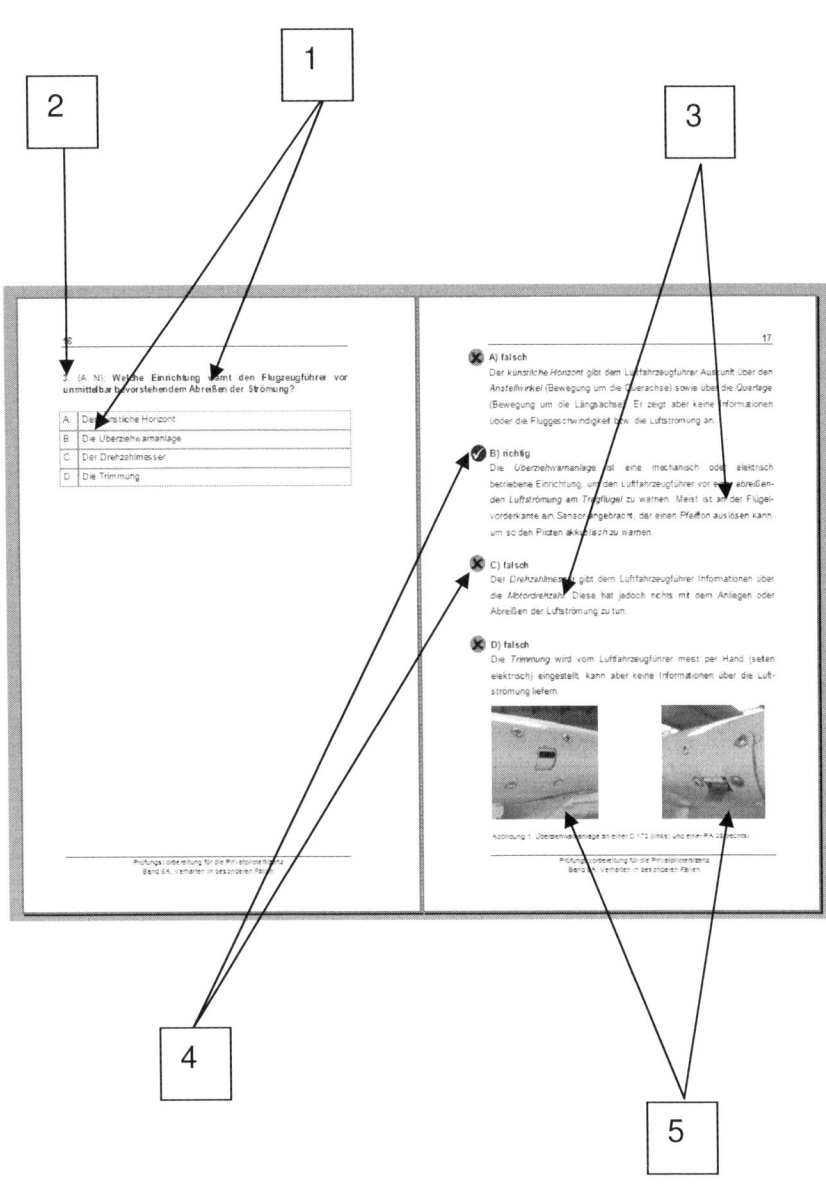

Inhaltsverzeichnis

8.2 Allgemeines Sprechfunkzeugnis für den Flugfunk (AZF)

8.2.1 AZF Teil I

1. Select the correct definition for "estimated time of arrival" in respect to IFR flights:

A)   The time at which it is estimated that the aircraft will arrive over the designated point, defined by reference to navigation aids, from which it is intended, that an instrument approach will be commenced.

B)   The time at which it is estimated that the aircraft will arrive over that designated point defined by reference to visual aids, from which it is intended, that an approach will be commenced.

C)   The time at which the aircraft will actually arrive over that designated point defined by reference to navigation aids, from which it is intended, that a visual approach will be commenced.

D)   In any case that time at which the aircraft will arrive over the aerodrome.

 **A) Correct**

The *"estimated time of arrival"* (*ETA*) is defined as the time at which it is estimated that the aircraft will *arrive* over the *designated point* (e.g., reporting point or airfield). The ETA is referred to *navigation aids*, from which it is intended, that an instrument approach will be commenced.

B) Wrong

The *ETA* in respect to *VFR* flights (visual flight rules) is referred to *visual aids*. In contrast, the ETA in respect to IFR flights (instrument flight rules) is referred to navigation aids.

C) Wrong

The time at which the aircraft will *actually arrive* over that designated point is called the *"actual time of arrival"* (*ATA*).

D) Wrong

The ETA is not only referred to *aerodromes*.

 Die ETA (Estimated Time of Arrival; geschätzte Ankunftszeit) ist die Zeit, zu der das Luftfahrzeug vermutlich über einem definierten Punkt ankommen wird.

2. What does the term "aeronautical station" mean?

A)   A land station in the aeronautical mobile service, in certain
        instances, may be located, for example, on board of a ship or on
        a platform at sea

B)   Any station established to exchange radiotelephony
        communications

C)   A station in the aeronautical telecommunication service located
        on land or on board of an aircraft to exchange radiotelephony
        communications

D)   A station forming part of the aeronautical telecommunication
        network

A) Correct

An aeronatutical station is a land station in the *aeronautical mobile service*, in certain instances, may be located, for example, on board of a *ship* or on a *platform* at sea.

 B) Wrong

An aeronautical station is *not* "*any station*" but more specifically a land (ship or platform) station in the aeronautical mobile service.

 C) Wrong

The notion "aeronautical station" adresses the *aeronautical mobile service*. Stations belonging to the aeronautical telecommunication service are called ground stations due to the fact that the aeronautical telecommunication service mainly deals with message transmission via networks on the ground.

 D) Wrong

It is not a station in the *telecommunication network* but in the aeronautical mobile service.

3. What does the term "blind transmission" mean?

A)   A transmission of information relating to air navigation that is not addressed to a specific station or stations

B)   A transmission of messages relating to en-route weather information which may affect the safety of aircraft operations that is not addressed to a specific station or stations

C)   A transmission where no reply is required from the receiving station

D)   A transmission from one station to another station in circumstances where two-way communication cannot be established but where it is believed that the called station is able to receive the transmission

A) Wrong

Transmissions that are not addressed to specific stations are called *general calls* and not blind transmissions.

 B) Wrong

Transmissions *relating to en-route weather information* which may affect the safety of aircraft operations are *safety messages* or *weather reports* but not blind transmissions.

 C) Wrong

A transmission where *no reply is required* may be simple transmissions, e.g. weather/wind information.

 D) Correct

A *blind transmission* is a transmission from one station to another station in circumstances *where two-way communication cannot be established* but where it is believed that the called station is able to receive the transmission.

Example:

Pilot: D-EASY transmitting blind right base runway 03, D-EASY transmitting blind right base runway 03

 Blindsendungen werden von einer Station zu einer anderen übertragen, wenn man davon ausgeht, dass die andere Station die Nachricht empfangen kann.

4. What does the term "broadcast" mean?

A)   A radiotelephony transmission from a ground station to aircraft in flight

B)   A transmission where no reply is required from the receiving station

C)   A transmission containing meteorological and operational information to aircraft engaged in flights over remote and oceanic areas out of range of VHF ground stations

D)   A transmission of information relating to air navigation that is not addressed to a specific station or stations

A) Wrong

It is important that "*broadcast*" transmissions are *not addressed* to a specific station. This point lacks in the answer.

 B) Wrong

It is important that "*broadcast*" transmissions are *not addressed* to a specific station. This point lacks in the answer.

 C) Wrong

It is important that "*broadcast*" transmissions are *not addressed* to a specific station. This point lacks in the answer.

 D) Correct

Broadcast is a *transmission of information* relating to air navigation that is *not addressed to a specific station or stations*. **Example (ATIS):**

This is Mannheim Airport, Information Alpha, Met Report Time 1620, LOC-DME approach runway 27, transition level 60, wind 240 degrees, 5 knots, visibility 9 kilometers, light rain, clouds scattered 3,000 feet, temperature 24, dewpoint 18, QNH 1018, NOSIG, Information Alpha out."

 Unter "Broadcast" versteht man die Übermittlung von Informationen in Bezug zur Navigation, die nicht an eine bestimmte Funkstelle gerichtet sind.

5. What does the term "air-ground communication" mean?

A)   Two-way communication between aircraft and stations or locations on the surface of the earth

B)   One-way communication from aircraft to stations or locations on the surface of the earth

C)   One-way communication from stations or locations on the surface of the earth

D)   Any communication from aircraft to ground stations requiring handling by the Aeronautical Fixed Telecommunication Network (AFTN)

 **A) Correct**

"*Air-ground communication*" is defined as a *two-way communication* between aircraft and stations or locations on the surface of the earth. **Examples:**

- Aircraft ↔ Tower
- Aircraft ↔ ATC
- FIS ↔ Aircraft

 **B) Wrong**

Air-ground communication is *two-way but not one-way communication.*

 **C) Wrong**

Air-ground communication is *two-way but not one-way communication.* In addition, two-way communication *includes communication between airplanes.*

 **D) Wrong**

Air-ground communication has nothing to do with the *Aeronautical Fixed Telecommunication Network* (AFTN).

 Unter "Luft-Boden-Kommunikation" (air-ground communication) versteht man eine Funkverbindung zwischen zwei Stationen (eine in der Luft und eine am Boden).

6. What does the term "expected approach time" mean?

A)   The time at which an arriving aircraft, upon reaching the radio aid serving the destination aerodrome, will commence the instrument approach procedure for a landing

B)   The time at which an arriving aircraft expects to arrive over the appropriate designated navigation aid serving the destination aerodrome

C)   The holding time over the radio facility from which the instrument approach procedure for a landing will be initiated

D)   The time at which ATC expects that an arriving aircraft, following a delay, will leave the holding point to complete its approach for a landing

A) Wrong

The time at which an arriving aircraft, upon reaching the radio aid serving the destination aerodrome, will commence the instrument approach procedure for a landing is called the ETA (*"estimated time of arrival"*).

 B) Wrong

The *"estimated time of arrival"* (ETA) is the time at which it is estimated that the aircraft will arrive over a designated point.

 C) Wrong

When flying *IFR enroute*, a pilot may be requested to hold over an *enroute reporting point* until a *specified time*. This time-point is called *"holding time"*.

The *holding point* may be a navigational aid (e.g. *VOR* or *NDB*) but also a specified *geographical location* in relation to which the position of the aircraft can be reported.

 D) Correct

The *"expected approach time"* (EAT) is the time issued by the ATC at which it is estimated to *begin with the approach*, i.e. the time at which the initial approach fix (IAF) being *overflown* or the aircraft will *leave the holding point/holding pattern* to complete its approach.

 Die EAT (expected approach time) ist die Zeit, zu der ein Luftfahrzeug nach einer Verspätung vorraussichtlich das IAF überfliegt bzw. die Warteschleife verlässt und mit dem Anflug beginnt.

7. What does the term "visual approach" mean?

A)   An approach executed by a VFR flight unable to maintain VMC

B)   A visual manoeuvre executed by an IFR flight when the weather conditions at the aerodrome of destination are equal or better than required VMC minima

C)   An approach by an IFR flight when either part or all of an instrument approach procedure is not completed and the approach is executed in visual reference to terrain

D)   An extension of an instrument approach procedure to bring an aircraft into position for landing on a runway which is not suitably located for straight-in approach

A) Wrong

VFR flights which are unable to maintain VMC are *forbidden*.

 B) Wrong

A visual approach is not a *visual manoeuvre*.

 C) Correct

A *visual approach* is an approach by an *IFR flight* when either *part or all* of an instrument approach procedure is not completed and the *approach is executed in visual reference to terrain*.

 D) Wrong

A visual approach is *not an extension of an instrument approach*.

 Ein "visual approach" ist ein IFR-Anflug, bei dem ein Teil bis zur Landung nach VFR geflogen wird.

8. What does the term "clearance limit" mean?

A)   The time of expiry of an air traffic control clearance

B)   The time at which an aircraft is given an air traffic control clearance

C)   A point to which an aircraft is granted an air traffic control clearance

D)   The time after which an air traffic control clearance will be automatically cancelled if the flight has not been commenced

A) Wrong

The "clearance limit" is *not a time but a physical point* to which an aircraft is granted an air traffic control clearance.

 B) Wrong

The "clearance limit" is *not a time but a physical point* to which an aircraft is granted an air traffic control clearance.

 C) Correct

The "*clearance limit*" is a *point to which an aircraft is granted an air traffic control clearance.*

 D) Wrong

The "clearance limit" is *not a time but a physical point* to which an aircraft is granted an air traffic control clearance.

 Das "Clearance limit" ist der Punkt, bis zu dem das Luftfahrzeug eine Freigabe erhalten hat.

9. What does the term "Automatic Terminal Information Service" mean?

A)   A service established to provide information concerning en-route weather phenomena which may effect the safety of an aircraft operation

B)   A service by which aircraft within a flight information region (FIR) are provided with current meteorological and operational information essential for the safety of air navigation

C)   The provision of current routine information to arriving and departing aircraft by means of continuous and repetitive broadcast throughout the day or a specific portion of the day

D)   A service which provides aircraft with weather reports relating to a specific number of aerodromes located within a flight information region (FIR)

A) Wrong

ATIS covers *airport weather* but not *en-route weather* information.

 B) Wrong

ATIS covers *airport weather* but not *weather information* in a flight information region (FIR).

 C) Correct

ATIS (Automatic Terminal Information Service) is the provision of current routine information to arriving and departing aircraft by means of continuous and repetitive broadcast throughout the day or a specific portion of the day. ATIS (Automatic Terminal Information Service) reports are *transmitted continuously on VOR-frequencies or separate ATIS-frequencies* and contain relevant actual weather information for take-off and landing. **Example (ATIS):**

> This is Mannheim Airport, Information Alpha, Met Report Time 1620, LOC-DME approach runway 27, transition level 60, wind 240 degrees, 5 knots, visibility 9 kilometers, light rain, clouds scattered 3,000 feet, temperature 24, dewpoint 18, QNH 1018, NOSIG, Information Alpha out.

 D) Wrong

VOLMET but not ATIS is a service which provides aircraft with weather reports relating to a specific number of aerodromes located within a *flight information region (FIR).*

10. What does the term "waypoint" mean?

A)   A defined position on an aerodrome used for the calibration of the inertial navigation system

B)   A specified geographical position used to define an area navigation route or the flight path of an aircraft employing area navigation

C)   A general term meaning the taxiway- and the runway-system of an international airport

D)   A signal indicating the direction of the runway-in-use

A) Wrong

A waypoint is a position on an area navigation route but *not on an airport.*

 B) Correct

A *waypoint* is a *specified geographical position* used to define an *area navigation route* or the *flight path* of an aircraft employing *area navigation.*

 C) Wrong

A waypoint is a position on an area navigation route and *has nothing to do with taxiways or runways on an airport.*

 D) Wrong

A waypoint is *not a signal but a specified geographical position.*

 Ein Waypoint ist eine Position, definiert durch Schnittpunkte von Luftverkehrsstraßen zur Area-Navigation.

11. What does the abbreviation "INS" mean?

A)   Inertial navigation system

B)   Instrument navigation system

C)   International NOTAM service

D)   International navigation service

 **A) Correct**

INS means "*Inertial Navigation System*". An inertial navigation system is a *navigation aid* that uses a *computer and motion sensors* to continuously track the *position*, *orientation*, and *velocity* (direction and speed of movement) of a vehicle without the need for external references.

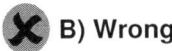 **B) Wrong**

An *Instrument navigation system* has no specific abbreviation.

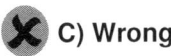 **C) Wrong**

The *International NOTAM service* has no abbreviation.

 **D) Wrong**

An *International navigation service* does not exist.

 Die Abkürzung INS bedeutet "Inertial Navigation System".

12. What does the abbreviation "TCAS" mean?

A)   Traffic alert and collision avoidance system

B)   Terminal control and advisory system

C)   Tower cabin alarm strop

D)   Track confirmation by automatic sources

 **A) Correct**

"*TCAS*" is the abbreviation for "*Traffic Alert and Collsion Avoidance System*". 'This system uses *radar and transponder* to identify other aircraft in the area and warns the pilot automatically to *change heading and/or altitude* to avoid a collision.

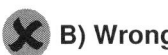 **B) Wrong**

Although "*Terminal control and advisory system*" sounds like something from aviation, it might be not a real thing.

 **C) Wrong**

"*Tower cabin alarm strop*" has nothing to do with aviation.

 **D) Wrong**

"*Track confirmation by automatic sources*" has nothing to do with aviation but may be a term used in computer science and electronics.

Die Abkürzung "TCAS" bedeutet "Traffic alert and Collision Avoidance System".

13. What does the abbreviation "SELCAL" mean?

A)   A system in which radiotelephony communication can be established between aircraft only

B)   A system which permits the selective calling of individual aircraft over radiotelephone channels linking a ground station with the aircraft

C)   A system in which radiotelephony communication between two stations can take place in both directions simultaneously

D)   A system provided for direct exchange of information between air traffic services (ATS) units

A) Wrong

SELCAL is not only a system for communication between aircraft but between *ground stations and aircraft* also.

 B) Correct

SELCAL (SELective CALling system) is a system which permits the selective calling of *individual aircraft over radiotelephone channels* linking a ground station with the aircraft. This communication is similar to normal radiotelephony communication (one-way communication) but addresses the communication *to a specific individual station only.*

 C) Wrong

With SELCAL radiotelephony communication is possible but *not in both directions simultaneously.*

 D) Wrong

The *air traffic service* has nothing to do with SELCAL.

 "SELCAL" ist ein Selektivrufsystem für die Luftfahrt.

14. What does the abbreviation "SSR" mean?

A)   Search and surveillance radar

B)   Secondary surveillance radar

C)   Surface strength of runway

D)   Standard snow report

A) Wrong

"SSR" is the abbreviation for "Secondary Surveillance Radar" but not for "*Search and Surveillance radar*". Such a device does not exist.

 B) Correct

"SSR" is the abbreviation for "*Secondary Surveillance Radar*". It works with secondary targets and not with primary targets of ordinary radar. SSR transmits radar pulses which are received by an airplane transponder. Thereafter the transponder sends a mode/code being visible on the controllers´ radar screen.

 C) Wrong

An abbreviation for "*Surface strength of runway*" does not exist.

 D) Wrong

An abbreviation for "*Standard snow report*" does not exist.

 SSR ist die Abkürzung für Secondary Surveillance Radar.

15. What does the abbreviation "RNAV" mean?

A.   Area navigation

B.   Radar aided navigation

C.   Route navigation

D.   Radio navigation

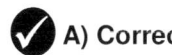 **A) Correct**

"*RNAV*" is the abbreviation for "*aRea NAVigation*" (R is the short version for Area). Classic navigation uses *radio navigation aids* (e.g. VOR or NDB) to define *aviation routes* and to facilitate navigation. In contrast, *RNAV is not dependend on these aids* and uses *waypoints* only for navigation. A *waypoint* is a *specified geographical position* used to define an *area navigation route* or the *flight path* of an aircraft employing *area navigation*. A *flight management system* (FMS) is essential for RNAV.

 **B) Wrong**

"*Radar aided navigation*" has no specified abbreviation.

 **C) Wrong**

"*Route navigation*" has no specified abbreviation.

 **D) Wrong**

"*Radio navigation*" has no specified abbreviation.

 RNAV ist die moderne Art der Navigation (Abkürzung für "Area Navigation").

16. What does the abbreviation "RVR" mean?

A)   Runway visual range

B)   Radar vectors requested

C)   Runway visibility report

D)   Recleared via route ...

 **A) Correct**

RVR is the abbreviation for "*Runway visual range*". This is the *maximum horizontal visibility on the runway* in the take-off and landing direction. RVR is essential for IFR-flights only.

B) Wrong

If radar vectors are requested, the pilot has to "*...request radar vectors*". For this request, an abbreviation is not allowed.

C) Wrong

Runway visibility reports may mention the RVR but RVR is not the abbreviation of those reports.

D) Wrong

"*Recleared via route...*" has no specific abbreviation and must not be abbreviated.

 RVR ist die Abkürzung für "Runway Visual Range".

17. What does the abbreviation "HJ" mean?

A)   Sunrise to sunset

B)   Sunset to sunrise

C)   No specific working hours

D)   Continuous day and night service

 **A) Correct**

"*HJ*" is the abbreviation for the time-period *sunrise to sunset*.

B) Wrong

The abbreviation for the time-period *sunset to sunrise* is "HN".

C) Wrong

"HX" means that there are *no specific working hours* defined.

D) Wrong

Continuous day and night service is abbreviated with "H24".

 „HJ" ist die Abkürzung für den Zeitraum Sonnenaufgang bis Sonnenuntergang.

18. QFE is the radio telephony Q-code to indicate:

A)   The atmospheric pressure referred to the highest fixed obstacle located on the surface of an aerodrome

B)   The atmospheric pressure referred to a point on the surface of the earth

C)   The altimeter sub-scale setting to obtain elevation when on the ground

D)   The atmospheric pressure at aerodrome elevation (or at runway threshold)

A) Wrong

The *atmospheric pressure referred to the highest fixed obstacle located on the surface of an aerodrome* has no specific abbreviation.

 B) Wrong

The *atmospheric pressure referred to a point on the surface of the earth* is the QFF (Table 1).

 C) Wrong

The *altimeter sub-scale setting to obtain elevation when on the ground* is QNH.

 D) Correct

QFE is the abbreviation (Q-code) for the *atmospheric pressure at aerodrome elevation* (or at runway threshold). This setting reports the present height.

Q-codes were originated in early aviation times, as radiotelephony communication has not been invented and only Morse codes did exist.

Abbreviation	Reference point
QNH	Mean Sea Level (MSL)
QFE	Ground (e.g. threshold at an airport)
QFF	Aerodrome Reference Point (ARP)

Table 1: Q-codes for altimeter setting and their reference points

 Der Luftdruck in Flugplatzhöhe oder an der Landebahnschwelle wird mit QFE abgekürzt!

19. QNH is the radio telephony Q-code to indicate:

A)   The atmospheric pressure at aerodrome elevation (or at runway
       threshold)

B)   The altimeter sub-scale setting to obtain elevation when on the
       ground

C)   The atmospheric pressure measured at the aerodrome reference
       point (ARP)

D)   The atmospheric pressure referred to the highest obstacle located
       on the surface of an aerodrome

A) Wrong

The abbreviation (Q-code) for the *atmospheric pressure at aerodrome elevation* (or at runway threshold) is the QFE. This setting reports the present height.

 B) Correct

The *altimeter sub-scale setting to obtain elevation when on the ground* is the QNH (Table 1, page 45).

 C) Wrong

The *atmospheric pressure referred to a point on the surface of the earth* (e.g. ARP) is the QFF.

 D) Wrong

The *atmospheric pressure referred to the highest fixed obstacle located on the surface of an aerodrome* has no specific abbreviation.

 Stellt man den Höhenmesser eines Luftfahrzeugs am Boden auf das QNH ein, zeigt dieser die Höhe des Flugplatzes über NN (MSL) an.

20. If you are requested to report your height, to which Q-code setting would you refer?

A)   QNH

B)   QFE

C)   QDM

D)   QBI

A) Wrong

QHN indicates the *altitude above MSL* but not the height.

 B) Correct

The abbreviation (Q-code) for the *atmospheric pressure at aerodrome elevation* (or at runway threshold) is the QFE. When setting the altimeter to QFE, it indicates the present height. During taxi it should indicate 0 ft and during flight the *height above the ground* (Table 1, page 45).

 C) Wrong

QDM is the *magnetic course* to a station but not an altimeter setting.

 D) Wrong

QBI is *not* a *used* abbreviation for altimeter settings.

 Stellt man am Höhenmesser das QFE ein, kann man die Höhe über Grund ablesen.

21. If you are requested to report your altitude, to which Q-code setting would you refer?

A)   QNH

B)   QFE

C)   QFF

D)   QNJ

 **A) Correct**

QHN refers to the air pressure at MSL and therefore indicates the *altitude above MSL* (Table 1, page 45).

B) Wrong

The abbreviation (Q-code) for the *atmospheric pressure at aerodrome elevation* (or at runway threshold) is the QFE. When setting the altimeter to QFE, it indicates the present height but not the altitude.

C) Wrong

The *atmospheric pressure referred to a point on the surface of the earth* (e.g. ARP) is the QFF. Therefore, it indicates a height.

D) Wrong

QNJ is not a frequently used abbreviation.

 Die Höhenmesser-Skalaeinstellung, die einem Luftfahrzeug die Flughöhe (Altitude) anzeigt, ist das QNH.

Prüfungsvorbereitung für die Privatpilotenlizenz
Band 8B: Allgemein gültiges Sprechfunkzeugnis (AZF)
2. Auflage 2009

22. What does "QDM" mean?

A)   Magnetic bearing to the station

B)   Magnetic bearing from the station

C)   True heading to the station (no wind)

D)   True bearing from the station

 **A) Correct**

QDM is the *magnetic bearing (heading) to the station* (Table 2).

 **B) Wrong**

The magnetic bearing from the station is the *QDR*.

 **C) Wrong**

The true heading to the station is the *QUJ*.

 **D) Wrong**

The true bearing from the station is the *QTE*.

Abbreviation	Meaning
QDM	Magnetic bearing to the station
QDR	Magnetic bearing from the station
QUJ	True heading to the station
QTE	True heading from the station

Table 2: Q-codes for navigation and their meaning

Prüfungsvorbereitung für die Privatpilotenlizenz
Band 8B: Allgemein gültiges Sprechfunkzeugnis (AZF)
2. Auflage 2009

23. What is the Q-code for "magnetic bearing to the station"?

A)   QDM

B)   QDR

C)   QNE

D)   QTE

 **A) Correct**

QDM is the *magnetic bearing (heading) to the station* (Table 2, page 53).

B) Wrong

The magnetic bearing from the station is the *QDR.*

C) Wrong

QNE is a Q-code for pressure altitudes but not for navigation.

D) Wrong

The true bearing from the station is the *QTE.*

 QDM ist der missweisende Kurs von einem Luftfahrzeug zu einer Station.

24. What does "QTE" mean?

A)   Magnetic bearing from the station

B)   True bearing from the station

C)   True heading to the station (no wind)

D)   Magnetic heading to the station

A) Wrong

The *magnetic bearing from the station* is the *QDR*. The magnetic bearings/headings are both QDM and QDR.

 **B) Correct**

The *true bearing from the station* is the *QTE* (Table 2, page 53). The Q-codes for true bearings/headings are both QTE and QUJ.

 C) Wrong

The true heading to the station is the *QUJ*.

 D) Wrong

QDM is the *magnetic bearing (heading) to the station*.

 Das QTE ist der rechtweisende Kurs von einer Station zu einem Luftfahrzeug.

25. What is the Q-code for "true bearing from the station"?

A)   QDR

B)   QDM

C)   QTE

D)   QFE

A) Wrong

The *magnetic bearing from the station* is the *QDR*. The magnetic bearings/headings are both QDM and QDR.

 B) Wrong

QDM is the *magnetic bearing (heading) to the station*.

 C) Correct

The *true bearing from the station* is the *QTE* (Table 2, page 53). The Q-codes for true bearings/headings are both QTE and QUJ.

 D) Wrong

The abbreviation (Q-code) for the *atmospheric pressure at aerodrome elevation* (or at runway threshold) is the QFE. When setting the altimeter to QFE, it indicates the present height but not the altitude.

 QDR±180°=QDM

26. What does "QDR" mean?

A)   Magnetic heading to the station (no wind)

B)   True bearing from the station

C)   True heading to the station

D)   Magnetic bearing from the station

A) Wrong

QDM is the *magnetic bearing (heading) to the station.*

 B) Wrong

The *true bearing from the station* is the *QTE* (Table 2, page 53). The Q-codes for true bearings/headings are both QTE and QUJ.

 C) Wrong

The true heading to the station is the *QUJ.*

 D) Correct

The *magnetic bearing from the station* is the *QDR.* The magnetic bearings/headings are both QDM and QDR.

Note: QDR±180°=QDM

 Das QDR ist der magnetische (missweisende) Kurs von einer Station zum Luftfahrzeug (Gegenkurs zum QDM).

27. What is the Q-code for "magnetic bearing from the station"?

A)   QDR

B)   QTE

C)   QDM

D)   QFE

 **A) Correct**

The *magnetic bearing from the station* is the *QDR* (Table 2, page 53). The magnetic bearings/headings are both QDM and QDR.

B) Wrong

The *true bearing from the station* is the *QTE*. The Q-codes for true bearings/headings are both QTE and QUJ.

C) Wrong

QDM is the *magnetic bearing (heading) to the station*.

D) Wrong

The abbreviation (Q-code) for the *atmospheric pressure at aerodrome elevation* (or at runway threshold) is the QFE (Table 1, page 45). When setting the altimeter to QFE, it indicates the present height but not the altitude.

 $QDM \pm 180° = QDR$

28. The message to a ground station on a landing site "PLEASE CALL A
TAXI CAB FOR US, WE WILL ARRIVE AT 1045" is...

A)   a flight regularity message

B)   a flight safety message

C)   an urgency message

D)   an unauthorized message

A) Wrong

Flight regularity messages are messages, e.g. for non-routine landings of aircraft. **Example:**

Pilot: D-GJOF diverting to EDRY

 B) Wrong

A *flight safety message* is used to control traffic, e.g. to avoid collsions. **Example:**

Tower: D-EASY make right turn, heading 180°

 C) Wrong

An *urgency message* is used, if an aircraft might be endangered, e.g. loss of oil-pressure indication, but still running engine. **Example:**

Pilot: PAN PAN PAN PAN PAN PAN D-EZOF one passenger has severe vomiting, request immediate landing

 **D) Correct**

Calling a taxi is definitely an unauthorized message!

 Die Meldung über die Ankunftszeit in Verbindung mit einem Wunsch nach einem Taxi ist keine erlaubte Meldung im beweglichen Flugfunkdienst.

29. Messages concerning non-routine landings of an aircraft are:

A)   Flight safety messages

B)   Urgency messages

C)   Flight regularity messages

D)   Unauthorized messages

A) Wrong

A *flight safety message* is used to control traffic, e.g. to avoid collsions.

Example:

Tower: D-EASY make right turn, heading 180°

 B) Wrong

An *urgency message* is used, if an aircraft might be endangered, e.g. loss of oil-pressure indication, but still running engine. **Example:**

Pilot: PAN PAN PAN PAN PAN PAN D-EZOF one passenger has severe vomiting, request immediate landing

 C) Correct

Messages concerning non-routine landings of an aircraft *are flight regularity messages.* **Example:**

Pilot: D-ABVA request refilling as soon as possible

 D) Wrong

Calling a taxi or asking for a hotel price is definitely an unauthorized message!

 Die Inhalte einer Flugbetriebsmeldung können beispielsweise die Anforderung von dringend benötigten Ersatzteilen oder die Meldung über nicht-planmäßige Landungen sein.

30. A message concerning aircraft parts and material urgently required is...

A)   an urgency message

B)   a flight safety message

C)   a flight regularity message

D)   a flight security message

A) Wrong

An *urgency message* is used, if an aircraft might be endangered, e.g. loss of oil-pressure indication, but still running engine. **Example:**

Pilot: PAN PAN PAN PAN PAN PAN D-EZOF one passenger has severe vomiting, request immediate landing

 B) Wrong

A *flight safety message* is used to control traffic, e.g. to avoid collsions. **Example:**

Tower: D-EASY make right turn, heading 180°

 C) Correct

Messages concerning non-routine landings of an aircraft are *flight regularity messages*. **Example:**

Pilot: D-EASY request new tires as soon as possible

 D) Wrong

Flight security messages do not exist.

31. Flight safety messages are...

A)   messages concerning non-routine landings

B)   messages concerning the safety of an aircraft, a vessel, any other
        vehicle or a person

C)   air traffic control messages

D)   messages relating to direction finding

A) Wrong

Messages concerning non-routine landings are *flight regularity messages.*

Example:

Pilot: D-EASY request new tires as soon as possible

 B) Wrong

Messages concerning the safety of an aircraft, a vessel, any other vehicle or a person are urgency messages. An *urgency message* is used, if an aircraft might be endangered, e.g. loss of oil-pressure indication, but still running engine. **Example:**

Pilot: PAN PAN PAN PAN PAN PAN D-EZOF one passenger has severe vomiting, request immediate landing

 **C) Correct**

A *flight safety message* is used to control traffic, e.g. to avoid collsions.

Example:

Tower: D-EASY make right turn, heading 180°

 D) Wrong

Messages relating to direction finding are *radio direction messages.*

Example:

Pilot: D-EASY request QDM

 Flugsicherheitsmeldungen sind beispielsweise Meldungen zur Flug-verkehrskontrolle.

32. A message concerning an aircraft being threatened by grave and imminent danger, requiring immediate assistance is called:

A)   Flight safety message

B)   Distress message

C)   Urgency message

D)   Class B message

A) Wrong

A *flight safety message* is used to control traffic, e.g. to avoid collsions.

Example:

Tower: D-EASY climb altitude 4,500 ft on heading 120°

 B) Correct

A message concerning an *aircraft being threatened by grave and imminent danger, requiring immediate assistance* is called distress message. **Example:**

Pilot: MAYDAY   MAYDAY   MAYDAY   D-EASY   engine   out, performing emergency landing 5 NM south of airfield

 C) Wrong

Messages concerning the safety of an aircraft, a vessel, any other vehicle or a person are urgency messages. An *urgency message* is used, if an aircraft or occupant might be endangered. **Example:**

Pilot: PAN PAN PAN PAN PAN PAN D-EZOF loss of oil-pressure indication, but engine is still running

 D) Wrong

Class B messages are not defined for aviation communication.

 Eine Notmeldung liegt dann vor, wenn ein Luftfahrzeug oder dessen Insassen akut bedroht sind und sofortige Hilfe benötigen.

33. Which of the messages listed below shall be handled by the aeronautical mobile service?

A)   Messages relating to direction finding

B)   Messages of airline operators

C)   Aeronautical security messages

D)   Aeronautical administrative messages

 **A) Correct**

Messages relating to direction finding are handled by the aeronautical mobile service (AMS), e.g. ATC. Additionally, the AMS handles messages concerning:

- Meteorological messages
- Flight safety messages
- Urgency messages
- Distress messages

Example:

Pilot: D-EASY request QDM

Tower: D-EASY QDM 135°

 B) Wrong

Messages of airline operators have to be transmitted by the airline operator but not by the the aeronautical mobile service.

 C) Wrong

Aeronautical security messages do not exist.

 D) Wrong

Aeronautical administrative messages may be published by postal mail.

34. Which of the messages listed below shall not be handled by the aeronautical mobile service?

A)   Meteorological messages

B)   Radio teletype messages

C)   Flight safety messages

D)   Urgency messages

A) Wrong

Meteorological messages are handled by the aeronautical mobile service (AMS). **Example:**

Pilot: D-EASY request windspeed and direction

Tower: D-EASY wind 120° / 6 kt

 B) Correct

Radio teletype messages are not handled by the aeronautical mobile service (AMS).

 C) Wrong

Flight safety messages are handled by the aeronautical mobile service (AMS).

Example:

Tower: D-EASY descent FL 90 on same heading

 D) Wrong

Urgency messages are handled by the aeronautical mobile service (AMS).

Example:

Pilot: D-EASY request landing as soon as possible, passenger with
                chest pain

Tower: D-EASY cleared to land runway 02, wind 030° / 5 kt

 Fernschreiben werden nicht vom Flugverkehrkontrolldienst verbreitet.

35. The priority of the instruction "TAXI TO HOLDING POSITION RUNWAY 05 VIA A" is...

A)   greater than "transmit for QDM"

B)   greater than "caution, construction work left of taxiway"

C)   less than "cleared to land"

D)   same as "line-up runway 07 and wait"

A) Wrong

The mentioned message is a *flight safety message* (#4; Table 3). "*Transmit for QDM*' is a direction finding message (#3). Thus, the presented answer is wrong.

 B) Wrong

The mentioned message is a *flight safety message* (#4; Table 3). "*Caution, construction work left of taxiway*' is also a flight safety message (#4). Thus, both messages have same but not greater priority.

 C) Wrong

The mentioned message is a *flight safety message* (#4; Table 3). "*Cleared to land*' is also a flight safety message (#4). Thus, both messages have same but not lower priority.

 D) Correct

The mentioned message is a *flight safety message* (#4; Table 3). "*Line-up runway 07 and wait*' is also a flight safety message. Thus, *both messages have same priority.*

Priority	Message
1	Distress message
2	Urgency message
3	Direction finding message
4	Flight safety message
5	Meteorological message
6	Flight regularity message
7	Radio teletype message

Table 3: Priority of messages

36. The message addressed to an Area Control Center "REQUEST RADAR VECTORS TO CIRCUMNAVIGATE ADVERSE WEATHER" is...

A)   a meteorological message

B)   an urgency message

C)   a message relating to direction finding

D)   a flight safety message

A) Wrong

The mentioned message is a *flight safety message* (#4;Table 3, page 79), since it requests radar vectors to navigate. Thus, *"meteorological message"* is wrong.

 B) Wrong

The mentioned message is a *flight safety message* (#4;Table 3, page 79), since it requests radar vectors to navigate. Thus, *"urgency message"* is wrong.

 C) Wrong

The mentioned message is a *flight safety message* (#4;Table 3, page 79), since it requests radar vectors to navigate. Thus, *"message relating to direction finding"* is wrong.

 D) Correct

The mentioned message is a *direction finding message* (#3;Table 3, page 79), since it requests radar vectors to navigate adverse weather.

 Bei einer Meldung mit der Radar-Vektoren zum Umfliegen von schlechtem Wetter angefordert werden, handelt es sich um eine Flugsicherheitsmeldung!

37. Air traffic control messages (clearances, instructions etc.) belong to the category of...

A)   class B messages

B)   flight regularity messages

C)   flight safety messages

D)   service message

A) Wrong

Class B messages are not defined for aviation communication.

 B) Wrong

Messages concerning non-routine landings of an aircraft are *flight regularity messages*. **Example:**

Pilot:　　　　　D-EASY request new tires as soon as possible

 C) Correct

A *flight safety message* is used to control air traffic, e.g. to avoid collisions, clearances or instructions. **Example:**

Tower:　　　　D-EASY climb altitude 4,500 ft on heading 120°

 D) Wrong

Service messages are not defined for aviation communication.

 Meldungen von ATC (z.B. Freigaben, Anweisungen etc.) gehören zu den Flugsicherheitsmeldungen.

38. The clearance "CLEARED FOR TAKE-OFF RUNWAY 03" is

A)   an urgency message

B)   a flight safety message

C)   an unauthorized message

D)   a flight regularity message

A) Wrong

Messages concerning the safety of an aircraft, a vessel, any other vehicle or a person are urgency messages. An *urgency message* is used, if an aircraft or occupant might be endangered. **Example:**

Pilot: PAN PAN PAN PAN PAN PAN D-EZOF loss of oil-pressure indication, but engine is still running

 B) Correct

"*CLEARED FOR TAKE-OFF RUNWAY 03*" is definitely a *flight safety message.*

 C) Wrong

For sure, "*CLEARED FOR TAKE-OFF RUNWAY 03*" is not an unauthorized message!

 D) Wrong

Messages concerning non-routine landings are *flight regularity messages.*
Example:

Pilot: D-EASY request new tires as soon as possible

 Freigaben von ATC sind Flugsicherheitsmeldungen!

39. The order of priority of the following messages in the aeronautical mobile service is...

A)   direction finding message, distress message, urgency message

B)   distress message, urgency message, direction finding message

C)   distress message, flight safety message, urgency message

D)   meteorological message, direction finding message, flight regularity message

A) Wrong

Priority is given in Table 3, page 79. *Therefore, the following priority is wrong:* direction finding message (#3), distress message (#1), and urgency message (#2).

 B) Correct

Priority is given in Table 3, page 79. Therefore, the following priority is right: *distress message (#1), urgency message (#2), and direction finding message (#3).*

 C) Wrong

Priority is given in Table 3, page 79. *Therefore, the following priority is wrong:* distress message (#1), flight safety message (#4), and urgency message (#2).

 D) Wrong

Priority is given in Table 3, page 79. *Therefore, the following priority is wrong:* meteorological message (#5), direction finding message (#3), and flight regularity message (#6)

40. The order of priority of the following messages in the aeronautical mobile service is...

A)   meteorological message, direction finding message, flight safety message

B)   flight regularity message, distress message, meteorological message

C)   flight safety message, meteorological message, flight regularity message

D)   flight safety message, direction finding message, urgency message

A) Wrong

Priority is given in Table 3, page 79. *Therefore, the following priority is wrong:* meteorological message (#5), direction finding message (#3), and flight safety message (#4).

 B) Wrong

Priority is given in Table 3, page 79. *Therefore, the following priority is wrong:* flight regularity message (#6), distress message (#1), and meteorological message (#5).

 C) Correct

Priority is given in Table 3, page 79. Therefore, the following priority is right: *flight safety message (#4), meteorological message (#5), and flight regularity message (#6).*

 D) Wrong

Priority is given in Table 3, page 79. *Therefore, the following priority is wrong:* flight safety message (#4), direction finding message (#3), and urgency message (#1).

41. The priority of the pilot's message "REQUEST QDM" is...

A)   less than "request climb flight level..."

B)   less than "descend flight level..."

C)   greater than "turn left heading..."

D)   same as "latest QNH 1018"

A) Wrong

"REQUEST QDM" is a *direction finding message* (#3; Table 3, page 79). "Request climb flight level..." is a *flights safety message* (#4). Thus, the answer is wrong.

 B) Wrong

"REQUEST QDM" is a *direction finding message* (#3; Table 3, page 79). "Descend flight level..." is a *flights safety message* (#4). Thus, the answer is wrong.

 C) Correct

"REQUEST QDM" is a *direction finding message* (#3; Table 3, page 79). "Turn left heading..." is a *flights safety message* (#4). Thus, the answer is *right*.

 D) Wrong

"REQUEST QDM" is a *direction finding message* (#3; Table 3, page 79). "Latest QNH 1018" is a *meteorological message* (#5). Thus, the answer is wrong.

 Bei der Meldung "Erbitte QDM" handelt es sich um eine Peilfunkmeldung mit der Priorität 3.

42. What is the correct way of spelling HBJYC ?

A)   Hotel Bravo Juliett India Kilo

B)   Hotel Bravo Juliett Yankee Charlie

C)   Hotel Bravo India Yankee Charlie

D)   Hotel Bravo India Victor Charlie

A) Wrong

"Y" is not "India" but "*Yankee*" and "C" is not "Kilo" but "*Charlie*".

 B) Correct

The correct spelling of the provided call sign is:

- *Hotel*
- *Bravo*
- *Juliett*
- *Yankee*
- *Charlie*

 C) Wrong

"J" is not "India" but "*Juliett*".

 D) Wrong

"J" is not "India" but "*Juliett*" and "Y" is not "Victor" but "*Yankee*".

 Insbesondere "I" (India), "J" (Juliett) und „Y" (Yankee) werden oftmals verwechselt.

43. What is the correct way of spelling FRI-VOR?

A)   Foxtrott Romeo Juliett - VOR

B)   Foxtrott Romeo India - VOR

C)   Fox Romeo Yankee - VOR

D)   Fox Romeo India - VOR

A) Wrong

"I" is not "Juliett" but "*India*".

 B) Correct

The correct spelling of the provided VOR station is:

- *Foxtrott*
- *Romeo*
- *India*
- *VOR*

 C) Wrong

"*Foxtrott*" (F) must not be abbreviated as "Fox". In addition, "I" is "*India*" but not "Yankee".

 D) Wrong

"*Foxtrott*" (F) must not be abbreviated as "Fox".

 „Foxtrott" (F) darf niemals durch „Fox" abgekürzt werden!

44. What is the correct way of transmitting the number 3500 ?

A)   Three five zero zero

B)   Three five hundred

C)   Three thousand five hundred

D)   Three five double "0"

A) Wrong

Hundreds and thousands and *their combination* may be spoken as a number and must not be spoken as separate digits. Thus, "Three five zero zero" is wrong.

 B) Wrong

Hundreds and thousands and *their combination* may be spoken as a number and must not be spoken as separate digits. Thus, "Three five hundred" is wrong.

 C) Correct

Hundreds and thousands and *their combination* may be spoken as a number and must not be spoken as separate digits. Therefore, "Three thousand five hundred" is the correct answer.

 D) Wrong

Hundreds and thousands and *their combination* may be spoken as a number and must not be spoken as separate digits. Thus, "Three five double "0"" is wrong.

 Ganze Hunderter und ganze Tausender, sowie deren Kombination müssen als ganze Zahl und nicht als einzelne Ziffern übermittelt werden.

45. What is the correct way of transmitting a QNH of 1001?

A)   QNH one zero zero one

B)   QNH one double "0" one

C)   QNH one thousand and one

D)   QNH one double zero one

A) Correct

Only *hundreds and thousands* and *their combination* may be spoken as a number and must not be spoken as separate digits. Since this applies not to "1001", the correct answer is "QNH one zero zero one".

 B) Wrong

Only *hundreds and thousands* and *their combination* may be spoken as a number and must not be spoken as separate digits. Thus "QNH one double "0" one" is wrong.

 C) Wrong

Only *hundreds and thousands* and *their combination* may be spoken as a number and must not be spoken as separate digits. Thus, "QNH one thousand and one" is wrong.

 D) Wrong

Only *hundreds and thousands* and *their combination* may be spoken as a number and must not be spoken as separate digits. Thus, "QNH one double zero one" is wrong.

46. What is the correct way of transmitting frequency 118.010 MHz (VHF channel spacing 8.33 kHz)?

A)   One eighteen decimal zero one

B)   One one eight decimal zero one zero

C)   One one eight point zero one zero

D)   One one eight decimal zero one

A) Wrong

Only *hundreds and thousands* and *their combination* may be spoken as a number and must not be spoken as separate digits. Therefore, "eighteen" is not allowed.

 B) Correct

For 8.33 kHz frequencies, *three digits after the decimal* have to be spoken. Therefore, "118.010" has to be spoken as "one one eight decimal zero one zero".

 C) Wrong

The "." is not "point" but "*decimal*".

 D) Wrong

For 8.33 kHz frequencies, *three* but not two *digits after the decimal* have to be spoken.

 Bei Frequenzen mit 8,33 kHz-Abstand müssen immer drei Ziffern gesprochen werden, da durch Nennung der ersten beiden Ziffern nach dem Komma die letzte Stelle nicht eindeutig definiert ist.

Prüfungsvorbereitung für die Privatpilotenlizenz
Band 8B: Allgemein gültiges Sprechfunkzeugnis (AZF)
2. Auflage 2009

47. What is the correct way of transmitting the number 13500 ?

A)   One three thousand five hundred

B)   One three five hundred

C)   One three five zero zero

D)   Thirteen thousand five hundred

A) Correct

Only *hundreds and thousands* and *their combination* may be spoken as a number and must not be spoken as separate digits. Therefore, "13500" has to be spoken as "one three thousand five hundred".

 B) Wrong

The word "*thousand*" is missing!

 C) Wrong

The words "*thousand*" and "*hundred*" are missing!

 D) Wrong

Only *hundreds and thousands* and *their combination* may be spoken as a number and must not be spoken as separate digits. This applies not to "13". Therefore, "*one three*" instead of "thirteen" is correct.

 „Dreizehntausend" wird als "eins drei tausend" übermittelt.

48. When transmitting time, which time system shall be used?

A)   Local time (LT), 24-hour clock

B)   Local time (LT), A.M. and P.M.

C)   No specific system, as only the minutes are normally required

D)   Co-ordinated universal time (UTC)

A) Wrong

Local time must not be used to transmit times.

 B) Wrong

Local time must not be used to transmit times.

 C) Wrong

The *time system* which has to be used is specifically defined (UTC, *universal time coordinated*).

 D) Correct

When transmitting times, UTC (*universal time coordinated*) has to be used.

 Bei der Übermittlung von Zeiten wird immer die Koordinierte Weltzeit (UTC) genutzt.

49. The time is 4:15 P.M. What is the correct way of transmitting this time if there is a possibility of confusion?

A)   Four fifteen P.M.

B)   Sixteen fifteen

C)   Four fifteen in the afternoon

D)   One six one five

A) Wrong

Only *hundreds and thousands* and *their combination* may be spoken as a number. But "15" has to be spoken as "*one five*" not as "fifteen".

 B) Wrong

Only *hundreds and thousands* and *their combination* may be spoken as a number. "16" as well as "15" have to be spoken as "*one six*" and "*one five*" but not as "sixteen" and "fifteen".

 C) Wrong

Only *hundreds and thousands* and *their combination* may be spoken as a number. But "15" has to be spoken as "*one five*" not as "fifteen". In addition, the suffix "*in the afternoon*" is not allowed.

 D) Correct

When *misunderstandings are not possible*, times may be transmitted as two digits, i.e. the minutes. But if there is a possibility of confusion, the 24 hour system has to be used. For the mentioned time (4:15 P.M. = 16:15 h), "*one six one five*" is the correct answer.

 Wenn Verwechselungsgefahr besteht, muss man Uhrzeiten vierstellig in einzelnen Ziffern übermitteln!

50. The time is 9:20 A.M. What is the correct way of transmitting this time if there is no possibility of confusion (same hour)?

A)   Two zero

B)   Twenty

C)   Two zero this hour

D)   Nine Twenty A.M.

A) Correct

When *misunderstandings are not possible*, times may be transmitted as two digits, i.e. the minutes.

Example: 10:35 a.m.   → "Three five"

 B) Wrong

Two digits (other than full hundreds or thousands or their combination) *have to be spoken separately*.

 C) Wrong

The appendix "*this hour*" must not be transmitted.

 D) Wrong

Two digits (other than full hundreds or thousands or their combination) have to be spoken separately. The appendix "*A.M.*" must not be transmitted.

 Sind Verwechslungen ausgeschlossen, können Uhrzeiten nur durch die beiden letzten Ziffern (Minuten) übermittelt werden.

51. An ATC unit providing air traffic control service to departing aircraft by means of surveillance radar has the call sign:

A)   DEPARTURE

B)   DELIVERY

C)   APPROACH

D)   CONTROL

A) Correct

"*DEPARTURE*" is the radiotelephony call sign for the aeronautical station indicating *approach control radar departures*.

 B) Wrong

"DELIVERY" is the call sign for *air traffic control* (ATC) assigning aircraft to RADAR after take-off.

 C) Wrong

"APPROACH" is the radiotelephony call sign for the aeronautical station providing *approach control service* (no radar service).

 D) Wrong

"CONTROL" is the radiotelephony call sign for the aeronautical station indicating an *area control centre* (no radar).

 "DEPARTURE" ist das Funkrufzeichen von ATC, die den radar-gestützten Abflug von Luftfahrzeugen leiten.

52. An ATC unit providing air traffic control service to enroute aircraft by means of surveillance radar has the call sign:

A)   CONTROL

B)   MONITOR

C)   DELIVERY

D)   RADAR

A) Wrong

"CONTROL" is the radiotelephony call sign for the aeronautical station indicating an *area control centre* (no radar).

 B) Wrong

The phrase "*MONITOR*" means the pilot will listen out on the given frequency or channel for further instructions or information. It is not a stations´ call sign.

 C) Wrong

"DELIVERY" is the call sign for *air traffic control* (ATC) assigning aircraft to RADAR after take-off.

 D) Correct

"RADAR" is the radiotelephony call sign of an ATC unit providing air traffic control service to enroute aircraft by means of surveillance radar.

 "RADAR" ist das Funkrufzeichen der Enroute-Luftverkehrskontrolle, welche die Radarführung ermöglicht.

Prüfungsvorbereitung für die Privatpilotenlizenz
Band 8B: Allgemein gültiges Sprechfunkzeugnis (AZF)
2. Auflage 2009

53. What is the radiotelephony call sign for the aeronautical station indicating aerodrome control?

A)   CONTROL

B)   AERODROME

C)   APRON

D)   TOWER

A) Wrong

"CONTROL" is the radiotelephony call sign for the aeronautical station indicating an *area control centre* (no radar).

 B) Wrong

"*AERODROME*" is not a used call sign.

 C) Wrong

"APRON" is the call sign for the *airport operator*, facilitating ground movements on the apron.

 D) Correct

"*TOWER*" is the radiotelephony call sign for the aeronautical station indicating aerodrome control, i.e. the control tower at controlled airfields or airports.

 Das Funkrufzeichen der Flugplatzkontrolle lautet "TOWER".

54. What is the radiotelephony call sign for the aeronautical station indicating flight information service?

A)   FLIGHT INFORMATION CENTRE

B)   FLIGHT CENTRE

C)   INFORMATION

D)   INFO

A) Wrong

The Flight Information Service has its office at the *Flight Information Centre (FIC)*, but this is not the call sign.

 B) Wrong

The Flight Information Service has its office at the Flight Information Centre (FIC), but *"FLIGHT CENTRE"* is not the call sign.

 C) Correct

"INFORMATION" is the call sign of the *Flight Information Service*.

 D) Wrong

"INFO" is the call sign of *non-controlled* (i.e. without air traffic control/TOWER) *airfields*.

 Der Fluginformationsdienst (FIS) besitzt das Rufzeichen "INFOR-MATION".

55. What is the radiotelephony call sign for the aeronautical station providing surface movement control of aircraft on the manoeuvring area?

A)   GROUND

B)   APPROACH

C)   TOWER

D)   CONTROL

 **A) Correct**

"*GROUND*" is the radiotelephony call sign for the aeronautical station providing *surface movement control* of aircraft on the manoeuvring area.

 **B) Wrong**

"APPROACH" is the radiotelephony call sign for the aeronautical station providing *approach control service* (no radar service).

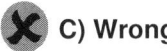 **C) Wrong**

"TOWER" is the radiotelephony call sign for the aeronautical station indicating *aerodrome control.*

D) Wrong

"CONTROL" is the radiotelephony call sign for the aeronautical station indicating an *area control centre* (no radar).

 Funkstellen mit dem Rufzeichen "GROUND" koordinieren die Rollbewegungen auf dem Boden.

56. What is the radiotelephony call sign for the aeronautical station providing approach control service (no radar service)?

A)   ARRIVAL

B)   RADAR

C)   CONTROL

D)   APPROACH

A) Wrong

"ARRIVAL" is the radiotelephony call sign for the aeronautical station indicating *approach control radar arrivals.*

 B) Wrong

"RADAR" is the radiotelephony call sign of an ATC unit providing *air traffic control service* to enroute aircraft by means of surveillance radar.

 C) Wrong

"CONTROL" is the radiotelephony call sign for the aeronautical station indicating an *area control centre* (no radar).

 D) Correct

"*APPROACH*" is the radiotelephony call sign for the aeronautical station providing *approach control service* (no radar service).

 Funkstellen mit dem Rufzeichen "APPROACH" führen den „Approach Control Service" durch.

57. What is the radiotelephony call sign for the aeronautical station indicating an area control centre (no radar)?

A)   CENTRE

B)   APPROACH

C)   CONTROL

D)   RADAR

A) Wrong

"*CENTRE*" is not used as a call sign.

 B) Wrong

"APPROACH" is the radiotelephony call sign for the aeronautical station providing *approach control service (no radar service)*.

 C) Correct

"*CONTROL*" is the radiotelephony call sign for the aeronautical station indicating an *area control centre* (no radar).

 D) Wrong

"RADAR" is the radiotelephony call sign of an ATC unit providing *air traffic control service* to enroute aircraft by means of surveillance radar.

 "CONTROL" ist ein Rufzeichen des ACC (Area Control Centre).

58. What is the radiotelephony call sign for the aeronautical station indicating approach control radar departures?

A)   CONTROL

B)   DEPARTURE

C)   RADAR

D)   APPROACH

A) Wrong

"CONTROL" is the radiotelephony call sign for the aeronautical station indicating an *area control centre* (no radar).

 B) Correct

"*DEPARTURE*" is the radiotelephony call sign for the aeronautical station indicating *approach control radar departures*.

 C) Wrong

"RADAR" is the radiotelephony call sign of an ATC unit providing *air traffic control service* to enroute aircraft by means of surveillance radar.

 D) Wrong

"APPROACH" is the radiotelephony call sign for the aeronautical station providing *approach control service (no radar service)*.

 Funkstellen mit dem Rufzeichen "DEPARTURE" führen Radar-Service für abfliegende Luftfahrzeuge durch.

59. What is the radiotelephony call sign for the aeronautical station indicating approach control radar arrivals?

     A)   ARRIVAL

     B)   APPROACH

     C)   RADAR

     D)   DIRECTOR

A) Correct

"*ARRIVAL*" is the radiotelephony call sign for the aeronatutical station indicating approach control radar arrivals.

 B) Wrong

"APPROACH" is the radiotelephony call sign for the aeronautical station providing *approach control service (no radar service).*

 C) Wrong

"RADAR" is the radiotelephony call sign of an ATC unit providing *air traffic control service* to enroute aircraft by means of surveillance radar.

 D) Wrong

"*DIRECTOR*" is not a used call sign.

 Funkstellen mit dem Rufzeichen "ARRIVAL" führen „radar approach control" für ankommende Luftfahrzeuge durch.

60. Which of the following abbreviated call signs of aircraft XYABC is correct?

A)   XYBC

B)   ABC

C)   BC

D)   XBC or XABC

A) Wrong

XYABC is abbreviated as *XBC* or *XABC* but not as XYBC.

 B) Wrong

XYABC is abbreviated as *XBC* or *XABC*. In this answer, "X" is missing.

 C) Wrong

Call signs may be abbreviated by using the *first and the two or three last characters*. In this answer, "X" is missing.

 D) Correct

This is the most suitable answer. Call signs may be abbreviated by using the *first and the two or three last characters*. In this case, XYABC is abbreviated as XBC or XABC.

Caution: Although XABC is considered as correct in this answer, it is considered wrong in question 61 (answer D)!

 Das Funkrufzeichen XYABC wird zu XBC abgekürzt.

61. Which of the following abbreviated call signs of Cherokee XY-ABC is correct?

A)   Cherokee XY-BC

B)   Cherokee BC or Cherokee ABC

C)   Cherokee X-BC

D)   Cherokee X-ABC

A) Wrong

Call signs may be abbreviated by using the *first and the two or three last characters*. In the mentioned situation, it is not allowed to *omit the "A" only*.

 B) Correct

Call signs may be abbreviated by using the *first and the two or three last characters*. The mentioned call sign is atypical for Germany, since the aircraft type is used seldomly as call sign. Additionally, the country locator is missing.

Example:

Pilot: Mannheim GROUND D-EASY

Ground: D-SY Mannheim GROUND

 C) Wrong

Call signs may be abbreviated by using the *first and the two or three last characters*. Therefore, the "X" has to be *omitted additionally*.

 D) Wrong

Call signs may be abbreviated by using the *first and the two or three last characters*.

Caution: Although X-ABC is considered as correct in question 60 (answer D), it is considered wrong in this answer!

 Funkrufzeichen werden mit dem ersten und den beiden letzten Buchstaben abgekürzt.

62. When shall an aircraft station use its abbreviated call sign?

A)   In low traffic

B)   After it has been addressed in this manner by the aeronautical ground station and no confusion is likely to result

C)   Upon initial call

D)   In dense traffic

A) Wrong

In *low traffic* it is not necessarily required to use the abbreviated call sign.

 B) Correct

An aircraft may only use the *abbreviated call sign* <u>after</u> it has been addressed in this manner *by the aeronautical ground station* and *no confusion* is likely to result. **Example:**

Pilot:	Langen RADAR D-EASY
Tower:	D-SY Langen RADAR identified, new heading 180°
Pilot:	D-SY new heading 180°

 C) Wrong

In the initial call, the *complete* and *non-abbreviated call sign* has to be used.

 D) Wrong

In *dense traffic* it may be useful to use the abbreviated call sign, but not by the pilot on his own decision.

 Ein Abgekürztes Funkrufkennzeichen darf erst dann benutzt werden, wenn es zuvor von der Bodenstation genutzt wurde.

63. What is the correct call sign of Fastair 345 in the initial call, if the aircraft has a maximum take-off mass of 136 tons or more?

A)   Heavy Fastair 345

B)   Fastair 345

C)   Fastair 345 widebody

D)   Fastair 345 heavy

A) Wrong

The word "heavy" has to be used immediately *after* the call sign.

 B) Wrong

Always when *establishing radiotelephony communications* with ATC and after each *frequency/channel change*, an aircraft >136 t should use the word "heavy" (in the wake-turbulence category) immediately after its call sign.

 C) Wrong

"*Heavy*" but not "*widebody*" is the word which has to be used immendiately after the call sign for aircrafts > 136 t MTOW.

 D) Correct

Always *when establishing radiotelephony communications* with ATC (initial call) and *after each frequency/channel change*, an aircraft >136 t should use the word "heavy" (in the wake-turbulence category) immediately after its call sign (Table 4, page 135). **Example:**

Pilot: Langen APPROACH, Air Berlin 48 heavy

Maximum Take-Off Weight (MTOW)	Wake-turbulence category for call sign
≤ 7 t	"light"
> 7 t and < 136 t	"medium"
≥ 136 t	"heavy"

Table 4: MTOW and wake-turbulence category for call sign

64. When shall an aircraft in the heavy-wake-turbulence category include the word "heavy" immediately after its call sign?

A)   Never

B)   In the initial call to the aerodrome control tower and the approach control unit

C)   always when establishing radiotelephony communications with ATC and after each frequency/channel change

D)   In all calls to the aerodrome control tower and the approach control unit

A) Wrong

Always when *establishing radiotelephony communications* with ATC and after each *frequency/channel change*, an aircraft >136 t should use the word "heavy" (in the wake-turbulence category) immediately after its call sign.

 B) Wrong

Not only for initial calls to the *aerodrome control tower* and the *approach control unit*, but in all initial calls the wake-turbulence category should be used.

 C) Correct

Always *when establishing radiotelephony communications* with ATC (initial call) and *after each frequency/channel change*, an aircraft >136 t should use the word "heavy" (in the wake-turbulence category) immediately after its call sign (Table 4, page 135). **Example:**

Pilot:　　　　Frankfurt RADAR, Lufthansa 367 heavy

 D) Wrong

It is not necessary to use the wake-turbulence category in all calls, but *only when establishing radiotelephony communications* (initial call) or *changing frequency*/channel.

65. When and by whom is the change of an aircrafts call sign in flight temporarily allowed?

A)   In the interest of safety to avoid confusion because of similar call signs by an ATC unit

B)   In case of a diversion to the alternate aerodrome by the pilot-in-command

C)   To faciliate subsequent radiotelephony communications by an aeronautical station

D)   When changing the destination airport during flight by the aircraft operator

A) Correct

In the *interest of safety* and to avoid confusion (because of similar call signs are used), the call sign of an aircraft in flight may be *temporarily changed by an ATC unit.*

 B) Wrong

The call sign of an aircraft may never be changed by *the pilot on his decision.*

 C) Wrong

To *facilitate subsequent radiotelephony communication*, the call sign of an aircraft may be used in the *abbreviated form*, but never be changed by the aeronautical station.

 D) Wrong

The call sign of an aircraft may never be changed by *the aircraft operator.*

 Das Funkrufzeichen darf nur dann gewechselt werden, wenn dies für die Sicherheit erforderlich ist (Vermeidung von Missverständnissen).

66. When may the name of the radiotelephony call sign of an aeronautical station/unit/service be ommitted?

A)   Never

B)   Only after the aeronautical station has used the abbreviated call sign

C)   In dense traffic during rush hours

D)   When satisfactory communication has been established and provided there will be no confusion

A) Wrong

The *radiotelephony call sign may be omitted* in some cases, e.g. when sufficient communication has been established.

 B) Wrong

There is *no relation between* the use of an *abbreviated* call sign and *omitting* the call sign.

 C) Wrong

In contrast, in *dense traffic rush hours* the complete call sign may be used to avoid *misunderstandings*!

 D) Correct

When *satisfactory communication* has been established and there will be *no confusion*, the radiotelephony call sign of an aeronautical station may be omitted. Actually, the call sign of ground stations usually is used only for the initial call and further on omitted. **Example:**

Pilot:	Mannheim TOWER D-EZOF, guten Tag
Turm:	D-EZOF Mannheim TOWER, guten Tag
Pilot:	Request taxi
Turm:	Taxi to holding point A runway 27

 Wenn bereits Funkverkehr aufgenommen wurde und eine Verwechslung ausgeschlossen ist, darf das Funkrufzeichen im Sprechfunkverkehr auch weggelassen werden.

67. What is the meaning of the phrase "CANCEL"?

A)   Annul the previous transmitted clearance

B)   Expect a new clearance shortly

C)   Consider that transmission as not sent

D)   A new flight plan has to be filed

A) Correct

"*CANCEL*" is the correct and official communication phrase to say "*Annul the previous transmitted clearance*". **Example:**

Tower: D-EFWM next reporting point Sierra2

Pilot: D-EFWM next reporting point Sierra2

Tower: D-EFWM CANCEL Sierra 2, next reporting point Echo

 B) Wrong

If a pilot has to *expect a new clearance shortly*, this will be transmitted in plain words by ATC.

 C) Wrong

"Disregard" would be the correct phrase to say "*Consider that transmission as not sent*".

 D) Wrong

If a *new flight plan has to be filed*, this may transmitted in plain words by ATC or the pilot.

 Die Sprechgruppe "CANCEL" macht eine zuvor übermittelte Meldung ungültig.

68. What is the meaning of the phrase "ACKNOWLEDGE"?

A)   I have received all of your last transmission

B)   Let me know that you have received and understood this
         message

C)   Repeat all or the following part of your last transmission

D)   My transmission is ended and I expect a response from you

A) Wrong

"Roger" would be the phrase to indicate "*I have received all of your last transmission*".

 B) Correct

"*ACKNOWLEDGE*" is the phrase to indicate "Let me know that you have received and understood this message".

 C) Wrong

"Say again" would be the phrase to indicate "*Repeat all or the following part of your last transmission*".

 D) Wrong

"Go ahead" would be the phrase to indicate "*My transmission is ended and I expect a response from you*".

 Mit der Sprechgruppe "ACKNOWLEDGE" fordert man die andere Funkstelle auf, den vollständigen Erhalt und Verständnis einer Nachricht zu bestätigen.

69. What does the phrase "ROGER" mean?

A)   A direct answer in the affirmative

B)   A direct answer in the negative

C)   I have received all of your last transmission

D)   Cleared for take-off or cleared to land

A) Wrong

The phrase "*affirmative*" does not exist in aviation communication (although it is frequently used, actually).

 B) Wrong

If something is negative, the phrase "*negative*" has to be used.

 C) Correct

The phrase "*ROGER*" means that a message has been *received completely*.

Example:

Tower: D-EASY wind 245 degrees, 8 knots

Pilot: D-EASY ROGER

 D) Wrong

"*Cleared for take-off or cleared to land*" must be used as a phrase. These clearances may not be abbreviated.

 Mit "ROGER" zeigt man an, dass man die letzte Meldung vollständig erhalten hat.

70. What does the phrase "STANDBY" mean?

A)   Continue on present heading and listen out

B)   Select STANDBY on the SSR transponder

C)   Wait and I will call you soon

D)   Permission granted for action proposed

A) Wrong

"*Continue on present heading and listen out*" is not an official phrase for communication.

 B) Wrong

If RADAR instructs aircraft XYABC "*XBC SQUAWK STANDBY*", the pilot should *switch the transponder-button to "standby"* (STBY or SBY, Figure 4, page 249). For this purpose the word "squawk" has to be added in the instruction!

 C) Correct

"*STANDBY*" is the correct phrase to indicate "Wait and I will call you".

 D) Wrong

"Cleared" would be the phrase to indicate "*Permission granted for action proposed*".

 Mit der Sprechgruppe "STANDBY" wird man angewiesen, auf weitere Informationen zu warten.

71. What does the phrase "READ BACK" mean?

A)   Let me know that you have received and understood this message

B)   Did you correctly receive this message?

C)   Repeat all, or the specified part, of this message back to me exactly as received

D)   Check and confirm with originator

A) Wrong

"Confirm" would be the phrase to say "*Let me know that you have received and understood this message*".

 B) Wrong

"Confirm" would be the correct phrase to say "*Did you correctly receive this message?*".

 C) Correct

"*READ BACK*" is the official and correct phrase to indicate "*Repeat all, or the specified part, of this message back to me exactly as received*".

 D) Wrong

"Check" would be the correct phrase to say "*Check and confirm with originator*".

 Mit der Sprechgruppe "READ BACK" zeigt man an, dass die andere Funkstelle eine Meldung wort-wörtlich wiederholen soll.

72. What does the phrase "CHECK" mean?

A)   Repeat your last transmission

B)   Did you correctly receive this message?

C)   Examine a system or procedure

D)   Consider that transmission as not sent

A) Wrong

"*SAY AGAIN*" is the official phrase to indicate *another station should repeat the last transmission* (e.g. if it was only understandable as a fragment). Although "SAY AGAIN" is slightly unpolite, it may be used by both air- and ground-stations.

 B) Wrong

"*Confirm*" would be the correct phrase to say "Did you correctly receive this message?".

 C) Correct

"*CHECK*" is the correct phrase for "Examine a system or procedure".

 D) Wrong

"*Disregard*" would be the correct phrase to say "Consider that transmission as not sent".

 "CHECK" ist die offizielle Sprechgruppe um der anderen Station mitzuteilen, dass sie ein Verfahren oder eine Übermittlung prüfen soll.

73. Which phrase shall be used if you want to say: "I should like to know ..." or "I wish to obtain ..."?

A)   Request

B)   Report

C)   Acknowledge

D)   Confirm

A) Correct

"*REQUEST*" is the official and correct phrase if the station wants to say: "I should like to know ..." or "I wish to obtain ...". **Example:**

Pilot: D-EASY request QDM

Tower: D-EASY QDM 180 degrees

 B) Wrong

"*REPORT*" is the official phrase to indicate that the other station should pass the following information.

 C) Wrong

"*Acknowledge*" is the phrase to indicate "Let me know that you have received and understood this message".

 D) Wrong

"*Confirm*" is the phrase to authenticate an already transmitted message or information.

 "REQUEST" wird genutzt, um die andere Funkstelle aufzufordern, eine bestimmt Information zu übermitteln.

Prüfungsvorbereitung für die Privatpilotenlizenz
Band 8B: Allgemein gültiges Sprechfunkzeugnis (AZF)
2. Auflage 2009

74. Which phrase shall be used if you want to say: "Pass me the following information ..."?

A)   Report

B)   Request

C)   Say again

D)   Check

A) Correct

"*REPORT*" is the official and correct phrase to indicate that the other station should pass the following information. **Example:**

Tower: D-EASY report field in sight

Pilot: D-EASY will report field in sight

 B) Wrong

"*REQUEST*" is the phrase if the station wants to say: "I should like to know ..." or "I wish to obtain ...".

 C) Wrong

"*SAY AGAIN*" is the official phrase to indicate *another station should repeat the last transmission* (e.g. if it was only understandable as a fragment). Although "SAY AGAIN" is slightly unpolite, it may be used by both air- and ground-stations.

 D) Wrong

"*CHECK*" would be the correct phrase for "Examine a system or procedure".

 "REPORT" ist die Sprechgruppe, mit der man die andere Funkstelle auffordert, eine bestimmte Information zu senden.

75. Which phrase shall be used to confirm that a message has been repeated correctly?

A)   That is right

B)   Affirm

C)   That is affirmative

D)   Correct

A) Wrong

"*That is right*" is not an official aviation communication phrase, since "right" may be misunderstandable with "right-left".

 B) Wrong

"*AFFIRM*" is the correct phrase to say "*yes*" in official aviation communication.

 C) Wrong

"*That is affirmative*" is not an official aviation communication phrase. Additionally, the phrase "*affirmative*" does not exist in aviation communication (although it is frequently used, actually).

 D) Correct

"*Correct*" is the official aviation communication phrase to indicate that a message has been repeated correctly.

 "CORRECT" ist die Sprechgruppe um anzuzeigen, dass eine Meldung richtig wiederholt wurde.

76. Which phrase shall be used if you want to say: "An error has been made in this transmission (or message indicated). The correct version is ..."?

A)   QNH 1017, correction QNH 1016

B)   QNH 1017, negative QNH 1016

C)   QNH 1017, negative 1016

D)   QNH 1017, negative I say again 1016

A) Correct

If an *error has been made* in a transmission (or message) one may use the phrase *"CORRECTION"* for indication. Additionally the correct information has to be repeated afterwards. **Example:**

Tower: D-EPIA QNH 1016

Pilot: D-EPIA QNH 1017 CORRECTION QHN 1016

 B) Wrong

"NEGATIVE" is the phrase to indicate *"No"* but not the phrase to correct falsely transmitted information.

 C) Wrong

"NEGATIVE" is the phrase to indicate *"No"* but not the phrase to correct falsely transmitted information.

 D) Wrong

"NEGATIVE" is the phrase to indicate *"No"* but not the phrase to correct falsely transmitted information.

 Hat man bei der Übermittlung einer Meldung einen Fehler gemacht, nutzt man zur Richtigstellung die Sprechgruppe "CORRECTION".

77. Which phrase shall be used if the repetition of an entire message is required?

A)   Repeat your message

B)   Say again

C)   What was your message

D)   Repeat your last transmission

A) Wrong

"*Repeat your message*" is not an official aviation communication phrase.

 B) Correct

"*SAY AGAIN*" is the official phrase to indicate *another station should repeat the last transmission* (e.g. if it was only understandable as a fragment). Although "SAY AGAIN" is slightly unpolite, it may be used by both air- and ground-stations.

 C) Wrong

"*What was your message*" is not an official aviation communication phrase.

 D) Wrong

"*Repeat your last transmission*" is not an official aviation communication phrase.

 "SAY AGAIN" ist die offizielle Sprechgruppe um anzuzeigen, dass eine Meldung wiederholt werden soll.

78. Which phrase shall be used if you want to say: "Consider that transmission as not sent"?

A)   Cancel my last message

B)   Disregard

C)   Forget it

D)   My last transmission is cancelled

A) Wrong

"*Cancel my last message*" is not an official aviation communication phrase.

 B) Correct

"*Disregard*" means that the pilot should not comply with a previously received instruction or to consider a transmission as not being sent.

 C) Wrong

"*Forget it*" is not an official aviation communication phrase. It may be thought but never be spoken.

 D) Wrong

"*My last transmission is cancelled*" is not an official aviation communication phrase.

 "DISREGARD" ist die offizielle Sprechgruppe, um anzuzeigen, dass eine bereits gesendete Information als nichtig betrachtet werden soll.

79. Which phrase shall be used if you want to say: "I understand your message and will comply with it"?

A)   Roger

B)   Will comply with your instruction

C)   Ok, will do it

D)   Wilco

A) Wrong

It is not recommended to use the phrase *"roger"* since misunderstandings may occur. In this situation using "roger" is not appropriate.

 B) Wrong

The pilot has to use the phrase *"WICLO"* (=*Will comply with your instruction*) but not the spoken words.

 C) Wrong

"Ok, will do it" is not an official aviation phrase.

 D) Correct

"WILCO" is the phrase to say: "*I understand your message and will comply with it*" ("Will comply with your instruction"). Additionally, some specific information or instructions have to be repeated word by word (e.g. headings, altitude, flight level, clearances etc.).

 Mit "WILCO" zeigt man an, dass man eine bestimmte Aufforderung verstanden hat und nach ihr handeln wird.

80. Which phrase shall be used if you want to say: "Yes"?

A)   Yes

B)   Roger

C)   Affirmative

D)   Affirm

A) Wrong

It is not allowed to use the word "*yes*" in aviation communication.

 B) Wrong

It is not recommended to use the phrase "*roger*" since misunderstandings may occur.

 C) Wrong

The phrase "*affirmative*" does not exist in aviation communication (although it is frequently used, actually).

 D) Correct

The correct phrase to "say yes" is "*AFFIRM*" in aviation communication.

 "YES" = "JA" = "AFFIRM"

81. What does the phrase "MONITOR" mean?

A)   Wait and I will call you

B)   Establish a radio contact with ...

C)   Examine a system or procedure

D)   Listen out on (frequency/channel)

A) Wrong

"*STANDBY*" would be the correct phrase for "Wait and I will call you".

 B) Wrong

"*CONTACT*" would be the correct phrase for "Establish a radio contact with ...".

 C) Wrong

"*CHECK*" would be the correct phrase for "Examine a system or procedure".

 D) Correct

The phrase "*MONITOR*" means the pilot will listen out on the given frequency or channel for further instructions or information.

 Mit der Sprechgruppe "MONITOR" wird man aufgefordert, eine bestimmt Frequenz für weitere Informationen / Anweisungen abzuhören.

82. What does the instruction: "FASTAIR 345 STANDBY 118.9 FOR TOWER" mean?

A)   Fastair 345 should change frequency to 118.9 on which aerodrome data are being broadcast

B)   Fastair 345 should contact TOWER on 118.9

C)   Fastair 345 should standby on the current frequency

D)   Fastair 345 should change frequency to 118.9 on which TOWER will initiate further communications

A) Wrong

Aerodrome data is only transmitted on ATIS frequencies, not on *TOWER frequencies*.

 B) Wrong

To contact another station, the phrase "*CONTACT*" will be used.

 C) Wrong

Due to a new frequency is reported, the *pilot has to change to the other frequency*.

 D) Correct

The given instruction means, that *Fastair 345 should change frequency to 118.900 MHz on which TOWER will initiate further communications*. Actually the frequency must be provided with three digits after the decimal!

83. Fastair 345 has been instructed to contact Stephenville ARRIVAL on frequency 118.0. What is the correct way to indicate it will follow this instruction?

A)   Changing over Fastair 345

B)   Fastair 345 118.0

C)   Changing to ARRIVAL Fastair 345

D)   Stephenville ARRIVAL Fastair 345

A) Wrong

When changing frequency, the *digits of the new frequency* have to be repeated!

 B) Correct

When changing frequency, the *digits of the new frequency* have to be repeated! In this situation, the answer is very short and abbreviated to a maximum extent. Actually the frequency must provided with three digits after the decimal! The best and correct response would be as follows:

Pilot: Fastair 345 will contact Stephenville ARRIVAL frequency
                 118.00

 C) Wrong

When changing frequency, the *digits of the new frequency* have to be repeated!

 D) Wrong

When changing frequency, the *digits of the new frequency* have to be repeated!

 Zur Bestätigung einer neuen Frequenz müssen die Ziffern derer unbedingt wiederholt werden!

84. Which phrase shall a pilot use if he receives an instruction from ATC which he cannot follow?

A)   Negative instruction

B)   Unable to comply

C)   Impossible to make it

D)   Disregard

A) Wrong

The phrase "*negative instruction*" does not exist for aviation communication.

 B) Correct

If a pilot has received an instruction from ATC (Air Traffic Control) which he cannot follow, he has to use the phrase "*unable to comply*".

 C) Wrong

The phrase "*impossible to make it*" does not exist for aviation communication.

 D) Wrong

"*Disregard*" means that the pilot should not comply with a previously received instruction.

 Mit der Sprechgruppe "unable to comply" zeigt man an, dass man einer bestimmten Anweisung nicht Folge leisten kann.

85. Which phrase should a pilot use to inform ATC that he is initiating a missed approach procedure?

A)   Missed approach

B)   Pulling up

C)   Overshooting

D)   Going around

A) Wrong

The pilot has to *perform a missed approach procedure*, but he is required to use the phrase *"going around"*.

 B) Wrong

"Pulling up" means to pull the aircraft up, but has nothing to do with a missed approach procedure.

 C) Wrong

"Overshooting" is not an existing phrase for communication.

 D) Correct

If the phrase *"go around"* is used, the aircraft has to go around immediately. This means, the aircraft has to carry out the *missed approach procedure*.

 Mit der Sprechgruppe "going around" zeigt der Pilot an, dass er durchstartet und der "Missed Approach Procedure" folgt.

86. An aircraft had initially been cleared to climb to FL 100. For separation purposes the aircraft has to be levelled off at FL 80 for a few minutes. ATC will give this instruction by using the phrase:

A)   Stop climb at FL 80

B)   Level off at FL 80

C)   Maintain FL 80

D)   Cleared FL 80

A) Correct

If a pilot has *temporarily to terminate his climb flight*, although initially cleared to an *higher flight level* (FL), ATC uses the phrase *"stop climb at FL ..."*.

 B) Wrong

The pilot is *leveling off at FL80*, but the correct phrase is "stop climb at FL ...".

 C) Wrong

"*Maintaining*" means proceed with flying at FL... but not to stop the climb flight at a specific FL.

 D) Wrong

In this situation, the pilot has *already been cleared to climb FL100*. Therefore, another clearance for FL80 is not allowed.

 Wenn man (obwohl man bereits eine Freigabe für eine höhere Flugfläche erhalten hat) zeitweise den Steigflug unterbrechen muss, ist die richtige Sprechgruppe "stop climb (flight) at FL ...".

Prüfungsvorbereitung für die Privatpilotenlizenz
Band 8B: Allgemein gültiges Sprechfunkzeugnis (AZF)
2. Auflage 2009

87. The instruction to set the transponder to mode A and C, code 0410 wil be phrased by ATC:

A)   SQUAWK ALPHA 0410 AND CHARLIE

B)   SQUAWK ALPHA AND CHARLIE ON 0410

C)   SQUAWK 0410

D)   SQUAWK MODE ALPHA CODE 0410 AND MODE CHARLIE

A) Wrong

It is not possible to squawk ALPHA and CHARLIE *simultaneously*.

 B) Wrong

It is not possible to squawk ALPHA and CHARLIE *simultaneously*.

 C) Correct

When a pilot is requested to squawk mode C, this is *not specifically reported by ATC* (in contrast to mode A).

 D) Wrong

It is not possible to squawk ALPHA and CHARLIE *simultaneously*.

 Wenn man den Transpondermode C nutzen soll, wird dies (im Gegensatz zu Mode A) von ATC nicht gesondert mitgeteilt.

88. What is the meaning of the phrase "CANCELLING MY IFR FLIGHT"?

A)   The pilot changes from IFR to VFR

B)   The pilot continues VFR and closes his flight plan

C)   The pilot indicates that his landing is assured and he will not
         submit the landing time

D)   The pilot closes his flight plan

A) Correct

If a pilot requests to *change from an IFR to a VFR* flight, he is required to use the phrase "CANCELLING". **Example:**

Pilot: D-EASY request to cancel IFR

 B) Wrong

The correct phrase *to close a flight plan* is the phrase "request to close my flight plan".

 C) Wrong

If the pilot is *landing at an uncontrolled airfield*, he must submit the landing time.

 D) Wrong

The pilot himself *cannot close the flight plan*. This obliges ATC only.

 Wenn der Wechsel der Flugregeln von IFR nach VFR erfolgt, wird dies mit der Sprechgruppe "CANCELLING…" angezeigt.

89. Which phrase is used by ATC if a position report over a compulsory reporting point is not required?

A)   CANCEL POSITION REPORT OVER ... (fix)

B)   NO POSITION REPORT OVER ... (fix)

C)   DO NOT REPORT OVER ... (fix)

D)   OMIT POSITION REPORT OVER ... (fix)

A) Wrong

Canelling an position report is not possible and may therefore not be used.

 B) Wrong

The presented answer "*no position report ...*" is not an official phrase and may therefore not be used.

 C) Wrong

The presented answer "*do not report over ...*" is not an official phrase and may therefore not be used.

 D) Correct

Sometimes a position report over a fix is not necessary or required. In this case, the pilot is instructed by ATC as follows: "OMIT POSITION REPORT OVER ... (fix)". Example:

Radar: D-EASY omit position report over ANEKI

 Muss man sich an einem Pflichtmeldepunkt ausnahmsweise nicht bei ATC melden, wird die Sprechgruppe "OMIT POSITION REPORT OVER ... (fix)" genutzt.

90. Permission to taxi to the runway in use for departure will be phrased:

A)   TAXI TO HOLDING POSITION RUNWAY ... VIA ...

B)   CLEARED TO RUNWAY ... VIA ...

C)   RUNWAY ... TAXI VIA ...

D)   TAXI VIA ... TO RUNWAY ...

A) Correct

Permission to taxi to the runway in use for departure will be phrased by "*TAXI TO HOLDING POSITION RUNWAY ... VIA ...*". **Example:**

Ground: D-EASY taxi to holding position A runway 25L via taxiways J, L, B and A.

Pilot: D-EASY will taxi to holding position A runway 25L via taxiways J, L, B and A.

 B) Wrong

"*CLEARED TO RUNWAY...*" is not an official phrase and may not be used. Additionally, the word "TO" may be *misunderstandable* ("two"?) and has to be omitted.

 C) Wrong

The presented answer is *incomplete* and may not be used. Additionally, the presented phrase words are in the wrong order and the mandatory words HOLDING POSITION are omitted.

 D) Wrong

The *holding position* has to be mentioned in the permission!

 Eine Rollfreigabe wird immer durch die Sprechgruppe "TAXI TO HOLDING POSITION RUNWAY ... VIA ..." kenntlich gemacht.

91. If a pilot may start climb/descend at his convenience, ATC will use the phrase:

A)   WHEN READY CLIMB/DESCEND TO FL ...

B)   CLIMB/DESCEND FL ...

C)   CLIMB/DESCEND FL ... AT ANY TIME

D)   CLIMB/DESCEND FL ... AT YOUR CONVENIENCE

A) Wrong

The presented answer may be *misunderstandable* with *"ready"* at the holding position. Therefore, it may not be used.

 B) Wrong

ATC (Air traffic control) will give a descent instruction by using the phrase *"DESCEND FL ..."*. This means, the pilot has to begin a descent flight to the mentioned flight level (FL) or altitude *immediately*.

 C) Wrong

The phrase *"AT ANY TIME"* does not mean the pilot may commence a climb/descent at his convenience. Therefore, the presented answer is incorrect.

 D) Correct

ATC (Air traffic control) will give a descent instruction by using the phrase *"DESCEND FL ... AT YOUR CONVENIENCE"*. This means, the pilot is allowed to begin a descent flight to the mentioned flight level (FL) or altitude *at his convenience*.

92. ATC will give a descent instruction by using the phrase:

A)   DESCEND FL ...

B)   MAINTAIN FL ...

C)   LEAVE FL ... FOR FL ...

D)   CLEARED FL ...

A) Correct

ATC (Air traffic control) will give a descent instruction by using the phrase "*DESCEND FL ...*". This means, the pilot is allowed to begin a descent flight to the mentioned flight level (FL) or altitude immediately. **Example:**

Radar: D-EEDL descend FL 90

Pilot: D-EEDL will descend FL 90

 B) Wrong

To "*maintain altitude... / maintain FL ...*" means, the pilot should stay at the present altitude/FL and should not climb or descend.

 C) Wrong

"*Leave FL ... for FL ...*" means, the pilot is allowed to climb or descend to the given altitude/FL.

 D) Wrong

To "*cleared altitude... / cleared FL ...*" means, the pilot is allowed to climb or descend to the given altitude/FL.

 Die Sprechgruppe für eine Freigabe zum Sinkflug lautet "DESCEND FL".

93. Permission to taxi to the take-off position will be phrased

A)   TAXI TO TAKE-OFF POSITION

B)   CLEARED INTO POSITION AND HOLD

C)   CONTINUE TO TAKE-OFF POSITION AND HOLD

D)   LINE UP RUNWAY ...

A) Wrong

The phrase "*TAXI TO TAKE-OFF POSITION*" does not exist in aviation communication because the word "take-off" may be misunderstood as „cleared for take-off".

 B) Wrong

The phrase "*CLEARED INTO POSITION AND HOLD*" does not exist in aviation communication. Additionally, the word "cleared" may only be used to confirm a clearance (e.g. for take-off or landing).

 C) Wrong

The phrase "*CONTINUE TO TAKE-OFF POSITION AND HOLD*" does not exist in aviation communication.

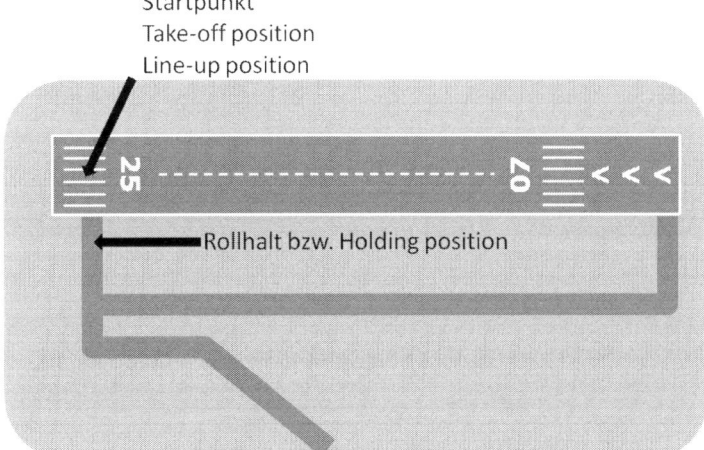 **D) Correct**

Permission to taxi to the take-off position will be phrased as "*LINE UP RUNWAY*". The *line up position* is the position on the *runway* from which the *take-off run is started* (Figure 1).

Figure 1: Line-up position, take-off position, and holding position.

Prüfungsvorbereitung für die Privatpilotenlizenz
Band 8B: Allgemein gültiges Sprechfunkzeugnis (AZF)
2. Auflage 2009

94. If requested by the control tower to report having crossed a runway, the pilot has to use the phrase:

A)   RUNWAY VACATED

B)   I HAVE LEFT THE RUNWAY

C)   I AM CLEAR OF THE RUNWAY

D)   I AM BEYOND THE RUNWAY

A) Correct

If requested by the control tower to report having crossed a runway, the pilot has to use the phrase *"RUNWAY VACATED"*. This indicates the runway is clear and may be used by another aircraft. Reportings should be *brief* and *not misunderstandable*.

 B) Wrong

The word "LEFT" (to leave) may be misunderstandable with "left" and "right". Therefore it may not be used. Additionally, reportings should be *brief* and *not misunderstandable*.

 C) Wrong

Reportings should be *brief* and *not misunderstandable*. This does not apply to the presented answer.

 D) Wrong

Reportings should be *brief* and *not misunderstandable*. This does not apply to the presented answer.

 Die Sprechgruppe, mit der man anzeigt, dass die Piste verlassen wurde, lautet "runway vacated".

95. In order to get a bearing, the ground station will request the pilot to...

A)   TRANSMIT FOR DF

B)   TRANSMIT FOR BEARING

C)   SQUAWK IDENT

D)   REPORT BEARING

A) Correct

During flight it is possible to obtain a *bearing (QDM)* from the ground station to have the *direct magnetic course* to the ground station. This bearing may be visible on the controllers´ ground screen, if the pilot *transmits for 2-3 seconds with his VHF radio.* Hereby, the correct instruction from the ground is: "*TRANSMIT FOR DF*" (DF=direction finder).

B) Wrong

The phrase "*transmit for bearing ...*" does not exist in aviation communication and is therefore no official phrase.

C) Wrong

RADAR instructs aircraft XYABC: "*XBC SQUAWK IDENT*". This means XBC shall operate (=*press*) the *IDENT-button* (Figure 3, page 247). The IDENT-button is on the transponder device at the cockpit panel. To be identified, the pilot should *press the button for approximately 2-3 seconds.*

D) Wrong

If the pilot is instructed to *report* anything he has to *answer* anything. In this situation the presented answer does not fit the question.

 Die richtige Sprechgruppe um dem Piloten anzuzeigen, dass er die Sprechtaste für die Bestimmung des QDM drücken soll, lautet „transmit for DF".

96. A pilot will be instructed to reselect the assigned transponder code A/S 6620 with the following phrase:

A)   SQUAWK AGAIN 6620

B)   SWITCH ON 6620

C)   RESET SQUAWK 6620

D)   CONFIRM SQUAWKING 6620

A) Wrong

The phrase "*squawk again* ..." does not exist in aviation communication and is therefore no official phrase.

 B) Wrong

The phrase "*switch on* ..." does not exist in aviation communication and is therefore no official phrase.

 C) Correct

A pilot will be instructed to _reselect_ the assigned *transponder code* A6620. This may appear if the pilot was temporarily instructed to use another mode/code. The correct phrase for this situation is: "_RESET ALPHA 6620_".

 D) Wrong

If the pilot should confirm squawking A6620, he is requested *to verify that he is already squawking* the mode/code A6620.

 Die Sprechgruppe zum Wiederherstellen eines entsprechenden Transpondercodes/-modes lautet "Reset A6620".

97. The prescribed phrase for obtaining permission to taxi to the runway for departure is:

A)   REQUEST TAXI CLEARANCE

B)   REQUEST PERMISSION TO TAXI

C)   WHAT IS MY TAXI CLEARANCE

D)   REQUEST TAXI

A) Wrong

To obtain a taxi clearance, the correct phrase is "request taxi" but not "*request taxi clearance*".

 B) Wrong

The words "*permission*" and "*to*" should be omitted. Therefore, this may not be the correct answer.

 C) Wrong

Requests should be *brief* ("request taxi") but *not polite* questions in aviation communication.

 D) Correct

To obtain a taxi clearance, the correct phrase is "*request taxi*". Requests should be *brief* ("request taxi") and *not misunderstandable*.

 Die richtige Sprechgruppe für eine Rollfreigabe lautet "request taxi".

98. ATC will give a climb instruction by using the phrase:

A)　　　MAINTAIN FL ...

B)　　　CLIMB FL ...

C)　　　LEAVE FL ... FOR FL ...

D)　　　CLEARED FL ...

A) Wrong

To "*maintain altitude... / maintain FL ...*" means, the pilot should stay at the present altitude/FL and should not climb or descend.

 B) Correct

The correct way for ATC to give a climb instruction is the phrase "*CLIMB FL ...*". This means, the pilot is *allowed to leave the present altitude/FL* for the given altitude/FL.

 C) Wrong

"*Leave FL ... for FL ...*" means, the pilot is allowed to climb or descend to the given altitude/FL.

 D) Wrong

To "*cleared altitude... / cleared FL ...*" means, the pilot is allowed to climb or descend to the given altitude/FL.

 Die richtige Sprechgruppe um in den Steigflug überzugehen lautet "CLIMB FL ...".

Prüfungsvorbereitung für die Privatpilotenlizenz
Band 8B: Allgemein gültiges Sprechfunkzeugnis (AZF)
2. Auflage 2009

99. A pilot intending to cancel his IFR flight plan and intending to continue VFR, shall use the phrase:

A)   REQUEST TO CLOSE MY FLIGHT PLAN

B)   Cancelling IFR or Cancelling IFR flight

C)   WILL CONTINUE VFR

D)   WILL CONTINUE VMC

A) Wrong

Closing the flight plan is not possible in this situation, since the aircraft is under IFR. Additionally, closing the flight plan does not induce cancelling IFR.

 B) Correct

The correct way to change flight rules from *IFR to VFR* is to *call ATC* (Air Traffic Control) and *cancel the IFR part.* **Example:**

Pilot: D-EASY cancelling IFR (flight)

 C) Wrong

"Will continue VFR" is a confirmation after closing the IFR part to proceed VFR in VMC.

 D) Wrong

"Will continue VMC" makes no sence since it does not mention the flight rules (VFR or IFR).

 Wird ein IFR Flug als VFR Flug fortgesetzt, wird dies mit der Sprechgruppe „request to cancel IFR flight" angefragt.

100. If a transponder does not transmit on mode C as expected, ATC will instruct the pilot to switch on mode C by using the phrase:

A)   SQUAWK ALTIMETER

B)   SQUAWK CHARLIE

C)   SQUAWK PRESSURE ALTITUDE

D)   TRANSMIT ON MODE CHARLIE

A) Wrong

"*Squawk altimeter*" is not an existing phrase in aviation communication.

 B) Correct

If a transponder does not transmit on mode C as expected, ATC will instruct the pilot to switch on mode C by using the phrase "*Squawk Charlie*". If mode alpha is requested, the instruction is "*Squawk alpha*".

 C) Wrong

"*Squawk pressure altitude*" is not an existing phrase in aviation communication.

 D) Wrong

"*Transmit on mode Charlie*" is not an existing phrase in aviation communication.

 Wenn man den Transponder-Mode C rasten soll, wird die Sprechgruppe "Squawk Charlie" genutzt.

101. The advice by a radar controller "TRAFFIC AT THREE O'CLOCK" means that the position of the mentioned traffic is...

A)   on the left side

B)   on the right side

C)   separated by three miles

D)   three miles ahead

A) Wrong

Three o´clock (Figure 2) is on the *right side of the clock face*; therefore "left" is incorrect.

 B) Correct

Three o´clock (Figure 2) is on the *right side of the clock face*.

 C) Wrong

If traffic would be *separated by three miles*, the *position* (ahead? left? right? etc.) is still *unclear*.

 D) Wrong

If the traffic would be *three miles ahead*, it would be reported as "traffic in 12 o´clock position, 3 miles".

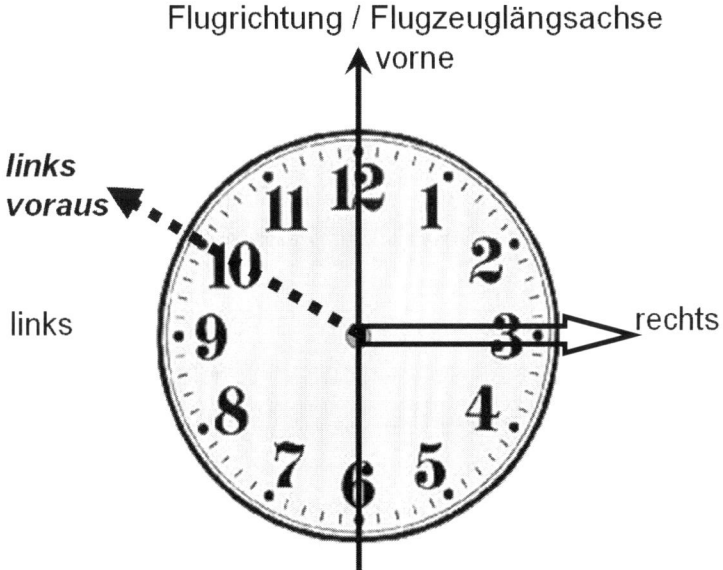

Figure 2: Directions according to clock positions

102. What is the correct procedure for the pilot to change from IFR flight to VFR flight?

A)   Call ATC and cancel IFR

B)   Call ATC and request to close the flight plan

C)   A change from IFR to VFR is not possible

D)   Call ATC and request clearance to proceed in accordance with the visual flight rules

A) Correct

The correct way to change flight rules from *IFR to VFR* is to *call ATC* (Air Traffic Control) and *cancel the IFR part.*

 B) Wrong

Closing the flight plan is not possible in this situation, since the aircraft is under IFR. Additionally, closing the flight plan does not induce cancelling IFR.

 C) Wrong

Instead, a *change from IFR to VFR is often necessary* (e.g. at smaller airfields and business aviation).

 D) Wrong

Requesting a clearance to proceed in accordance with VFR does not cancel the IFR part!

 Bei Wechsel der Flugregeln (IFR → VFR), muss der IFR-Teil aufgehoben werden!

103. What is the meaning of the phrase "CLEARANCE EXPIRES AT 1025"?

A)   The clearance is void if the aircraft departs before 1025

B)   The pilot shall stand by and ask for clearance again at 1025

C)   The pilot shall start engines not later than 1025

D)   The clearance is void if the aircraft is not airborne by or at 1025

A) Wrong

"*CLEARANCE EXPIRES AT 1025*" means that that clearance is void if the aircraft departs <u>after</u> 1025.

 B) Wrong

If the pilot shall *ask again for a clearance*, this is requested by ATC.

 C) Wrong

The answer is difficult. If the towed would have told "*clearance (for start-up) expires at 1025*", the answer could be right!

 D) Correct

"*CLEARANCE EXPIRES AT 1025*" means that that clearance is void if *the aircraft is not airborne by or at 1025*.

 Wenn eine Freigabe nur eine bestimmte Zeitdauer gültig ist, wird dies mit der Sprechgruppe "Clearance expires at…" mitgeteilt.

104. When shall the phrase "TAKE-OFF" be used by a pilot?

A)   Never, it is used only by the control tower

B)   To inform TOWER when ready for departure

C)   Only when the aircraft has already moved onto the active runway

D)   To acknowledge take-off clearance

A) Wrong

The phrase "*take-off*" may be used and must be by the pilot to confirm the take-off clearance.

 B) Wrong

If an aircraft is *prepared for take-off* (i.e. all items on the check-list done), the pilot may use the phrase "*ready*" or "*ready for take-off*".

 C) Wrong

The phrase "*take-off*" may be used and must be by the pilot to confirm the take-off clearance. This depends not on the *aircraft position*.

 D) Correct

The word "*take-off*" should not be used by the pilot to circumnavigate *misunderstandings* prior to the ATC clearance "*cleared for take-off*".

 Die Sprechgruppe "Take-off" darf nur als Bestätigung der Start-freigabe benutzt werden.

105. How shall a pilot inform the control tower that he is prepared for take-off?

A)   Ready for take-off

B)   Ready for departure or ready

C)   Ready to line-up

D)   Ready to go

A) Wrong

The word "*take-off*" should not be used by the pilot to circumnavigate *misunderstandings* prior to the ATC clearance "cleared for take-off".

 B) Correct

If an aircraft is *prepared for take-off* (i.e. all items on the check-list done), the pilot may use the phrase "*ready*" or "*ready for take-off*". **Example:**

Pilot: D-EZOF ready for departure

Tower: D-EZOF cleared for take-off runway 27

 C) Wrong

Lining-up means that the aircraft taxis to the line-up point on the rundway.

 D) Wrong

"*Ready to go*" is not official phrase for communication.

 Die Medung, dass ein Luftfahrzeug abflugbereit ist, lautet "ready (for departure).

106. How shall a pilot inform the control tower that he has to abandon the take-off manoeuvre?

A)   Abandoning take-off

B)   Stopping

C)   Aborting take-off

D)   Cancelling take-off

A) Wrong

"*Abdonning take-off*" is not an official phrase for communication.

 B) Correct

If a pilot has to *abdon (stop, cancel) the take-off manoeuvre*, he has to inform the control tower with the phrase "*stopping*".

 C) Wrong

"*Aborting take-off*" is not an official phrase for communication.

 D) Wrong

"*Cancelling take-off*" is not an official phrase for communication.

 Wir der Startvorgang abgebrochen, muss die Sprechgruppe "Stopping" genutzt werden.

107. How shall a pilot inform the control tower that he has to perform a missed approach?

A)   Overshooting

B)   Will make another approach

C)   Going around

D)   Pulling up

A) Wrong

"*Overshooting*" is not an existing phrase for communication.

 B) Wrong

If a pilot decides to make another approach, e.g. during VFR flight, he may use the prase "… *will make another approach*". This applies not for an IFR missed approach procedure.

 C) Correct

If the phrase "*go around*" is used, the airctaft has to go around immediately. This means, the aircraft has to carry out the *missed approach procedure*.

 D) Wrong

"*Pulling up*" means to pull the aircraft up, but has nothing to do woth a missed approach procedure.

 Muss ein "missed approach procedure" geflogen werden, nutzt der Pilot die Sprechgruppe "Go around".

108. What does the phrase "GO AROUND" mean?

A)   Overtake the aircraft ahead

B)   Carry out a missed approach

C)   Make a 360° turn

D)   Proceed with your message

A) Wrong

For *overtaking another aircraft*, no specific phrase does exist.

 B) Correct

If the phrase *"go around"* is used, the airctaft has to go around immediately. This means, the aircraft has to carry out the *missed approach procedure*.

 C) Wrong

If a pilot is instructed to *"orbit"*, he is requested to *make 360° turns*. Additionally, the direction (right/left) is reported.

 D) Wrong

If a pilot should proceed with his message, he is requested to "... *proceed (with your message)* ..."

 Bei der Sprechgruppe "Go around" muss sofort durchgestartet werden!

109. What does the phrase "ORBIT RIGHT" mean?

A)   Turn right to avoid other traffic

B)   Make 360° turns to the right

C)   Right-hand circuits are in use

D)   Leave the runway to the right

A) Wrong

If an aircraft should *turn right to avoid other traffic*, the pilot is instructed as follows: "D-EASY turn right immediately to avoid other traffic".

 B) Correct

If a pilot is instructed to "*orbit*", he is requested to *make 360° turns*. Additionally, the direction (right/left) is reported. **Example:**

Tower: D-EEDL orbit right

Pilot: D-EEDL will orbit right

 C) Wrong

If right-hand circuits are in use, the pilot is instructed to "... *make right-hand circuits*".

 D) Wrong

If a pilot should leave the rundway to the right, he is instructed to "*vacate right*".

 "Orbit" bedeutet, dass man Vollkreise (360°) fliegen soll.

110. What does the phrase "VACATE RUNWAY IMMEDIATELY" mean?

A)   Give way to aircraft from the left

B)   Turn left to leave the runway

C)   Hold position on the left side of the runway

D)   Clear the runway immediately

A) Wrong

The correct phrase to give way to an aircraft from the left is: "D-EASY *hold short before...* and proceed after..."

 B) Wrong

"VACATE RUNWAY" means to leave the runway but it does not mean to turn left.

 C) Wrong

If the pilot should hold position, he is instructed to "... *hold position* ...".

 D) Correct

"*VACATE RUNWAY IMMEDIATELY*" means that an aircraft should *leave (or clear) the runway* (in use) immediately following a further instruction.

.

 "Vacate" bedeutet verlassen!

111. What is the correct way for the pilot to acknowledge that ATIS information Golf has been received?

A)   Information Golf

B)   Weather Golf received

C)   We have the information

D)   We have the ATIS Golf

A) Correct

Having receipt an ATIS (*Aerodrome Terminal Information Service* or *Automatic Terminal Information Service*) information, may be simply confirmed by *naming the character* in the phrase "*.... received*". Example:

Pilot: D-EASY golf received

 B) Wrong

Having receipt an *ATIS information*, may be simply confirmed by naming the character in the phrase "information .... received" not "*weather .... received*".

 C) Wrong

Receiving an *ATIS information*, must be confirmed by naming the character, not only "*we have the information*".

 D) Wrong

Having receipt an *ATIS information*, may be simply confirmed by naming the character in the phrase "*information .... received*".

 Die korrekte Bestätigung einer ATIS-Meldung lautet: "<<Buchstabe>> received".

Prüfungsvorbereitung für die Privatpilotenlizenz
Band 8B: Allgemein gültiges Sprechfunkzeugnis (AZF)
2. Auflage 2009

112. What is the correct way to transmit and read back frequency/ channel 120.375 (VHF channel spacing 25 KHz)?

A)   One two zero decimal three seven

B)   One twenty decimal three seven

C)   One two zero decimal three seven five

D)   One two zero three seven

A) Wrong

When using *25 kHz channel spacing*, frequencies have to be transmitted as follows: *Three digits before* and *three digits after* decimal.

 B) Wrong

Only *hundreds and thousands* may be spoken as number, but "120" has to be spoken as separate digits: "One Two Zero".

 C) Correct

When using 25 kHz channel spacing, frequencies have to be transmitted as follows: *Three digits before* and *three digits after* decimal. **Example:**

Tower: D-EASY Contact Langen radar frequency 119.150

Pilot: D-EASY will contact Langen radar frequency 119.150

 D) Wrong

The *decimal* („.") has to be spoken, also.

 Frequenzen werden dreistellig vor und nach dem Komma übermittelt.

113. Which elements of instructions or information shall be read back?

A)   Runway-in-use, visibility, surface wind, heading instructtions, altimeter settings

B)   Surface wind, visibility, ground temperature, runway-in-use, altimeter settings, heading and speed instructions

C)   Runway-in-use, altimeter settings, SSR codes, level instructions, heading and speed instructions

D)   Surface wind, runway-in-use, altimeter settings, level instructions, SSR codes

A) Wrong

Meteorological information (e.g. visibility or wind) need not to be repeated!

 B) Wrong

Meteorological information (e.g. visibility, temperature or wind) need not to be repeated!

 C) Correct

The following elements have to be read back (repeated) in all cases:

- *Runway-in-use,*
- *altimeter settings,*
- *SSR (Transponder) codes,*
- *level instructions,*
- *heading and*
- *speed instructions*

 D) Wrong

Meteorological information (e.g. surface wind) need not to be repeated!

 Die aktuelle Piste, Einstellungen des Höhenmessers, Transponder codes, Flughöhe/Flight level, sowie Anweisungen über Kurs oder Geschwindigkeiten müssen immer wiederholt werden!

114. Shall an ATC route clearance be read back?

A)   No, if the ATC route clerance is transmitted in a published form
      (e.g. Standard Instrument Departure Route/SID)

B)   Yes, unless authorized otherwise by the ATS authority concerned

C)   No, if the communication channel is overleaded

D)   No, if the content of the ATC clearance is clear and no confusion
      is likely to arise

A) Wrong

Clearances are *not published*, but transmitted verbally.

 B) Correct

An ATC clearance has to be *read back in all cases*!

 C) Wrong

An ATC clearance has to be *read back in all cases*!

 D) Wrong

An ATC clearance has to be *read back in all cases*!

 Eine ATC clearance muss immer wiederholt werden!

115. An aircraft is instructed to hold short of the runway-in-use. What is the correct phraseology to indicate it will follow this instruction?

A)   Roger

B)   Holding short

C)   Will stop before

D)   Wilco

A) Wrong

In reality, it is very important to avoid misunderstandings in respect to taxiing. Therefore, confirming this instruction with "*roger*" is not sufficient.

 B) Correct

In reality, it is very important to avoid misunderstandings in respect to taxiing. Therefore, the instruction has to be repeated literally with "*holding short*". Caution: Actually, the *call sign* of the aircraft has to be transmitted also!

 C) Wrong

"*Will stop before*" is not an official phrase for communication.

 D) Wrong

In reality, it is very important to avoid misunderstandings in respect to taxiing. Therefore, confirming this instruction with "*wilco*" is not sufficient.

 Wichtige Anweisungen, z.B. zum Anhalten, müssen wörtlich wiederholt werden!

Prüfungsvorbereitung für die Privatpilotenlizenz
Band 8B: Allgemein gültiges Sprechfunkzeugnis (AZF)
2. Auflage 2009

116. Cherokee XYABC receives the following instruction: "XBC CLIMB STRAIGHT AHEAD UNTIL ALTITUDE 2500 FEET BEFORE TURNING RIGHT, WIND 270/6 KNOTS, CLEARED FOR TAKE-OFF". What is the correct readback?

A) Wilco, cleared for take-off, XBC

B) Right turn after 2500, roger, XBC

C) Straight ahead, 2500 feet right turn, wind west 6 knots, cleared for take-off, XBC

D) XBC straight ahead, at altitude 2500 feet right turn, cleared for take-off

A) Wrong

Instead of "WILCO", the pilot has to *repeat the complete clearance.*

 B) Wrong

Additionally, the pilot has to *read back "cleared for take-off".*

 C) Wrong

Reading back wind information is not necessary. But, the *call sign* should be reported at the beginning of the message.

 D) Correct

From the given information, *course instructions, altitudes/flight levels, and specific clearances* have to be repeated. Therefore, the correct read back is: XBC straight ahead, at altitude 2500 feet right turn, cleared for take-off.

Caution: XBC may only be used, if previously used by ATC!

 Freigaben, Kursanweisungen und Flughöhen müssen zwingend wörtlich wiederholt werden!

117. What does the phrase "SQUAWK 1234" mean?

A)   Give a short count for DF (direction finder)

B)   Make a test transmission on 123.4 MHz

C)   Standby on frequency 123.4 MHz

D)   Select Mode A and C / Code 1234 at the SSR transponder

A) Wrong

The correct phrase for a *short count for DF* is "TRANSMIT FOR DF".

 B) Wrong

The correct phrase for a *test transmission* is "MAKE A TEST TRANSMISSION ON FREQUENCY 123.4".

 C) Wrong

The correct phrase for *standby* is "STANDBY (ON THIS FREQUENCY)".

 D) Correct

A pilot receiving the phrase *"SQUAWK 1234"* should select Mode A and C / Code 1234 at the SSR transponder.

 Die richtige Sprechgruppe zum Rasten des Transponders mit einem bestimmten Code/Mode lautet „SQUAWK 1234".

118. RADAR informs aircraft XYABC: "XBC identified". What does this mean?

A)   XBC is not visible on the radar screen

B)   XBC should perform an identification turn

C)   Radar identification has been achieved

D)   XBC should operate the IDENT button

A) Wrong

If an aircraft is *not visible on the radar screen*, RADAR informs the aircraft XYABC: "XBC not identified".

 B) Wrong

If an identification turn should be performed, RADAR instructs the pilot: "*XBC perform identification turn of more than 20 degrees*".

 C) Correct

If *radar identification* has achieved, RADAR reports aircraft XYABC "XBC identified" or "XBC radar contact".

 D) Wrong

RADAR instructs aircraft XYABC: "*XBC SQUAWK IDENT*". This means XBC shall operate (=*press*) the *IDENT-button* (Figure 3, page 247). The IDENT-button is on the transponder device at the cockpit panel. To be identified, the pilot should *press the button for approximately 2-3 seconds*.

 Ist ein Luftfahrzeug vom Controller identifiziert, erhält der Luft-fahrzeugführer (z.B. von RADAR) die Meldung „XBC identified".

119. RADAR instructs aircraft XYABC: "XBC SQUAWK IDENT". What does this mean?

A)   Radar identification has been achieved by correlating an observed radar blip with aircraft XY-ABC

B)   XBC shall operate the IDENT-button (operation of the SPI feature)

C)   XBC should perform an identification turn of at least 20 degrees

D)   XBC shall reselect his assigned mode and code

A) Wrong

If *radar identification* has achieved, RADAR reports aircraft XYABC "XBC identified" or "XBC radar contact".

 B) Correct

RADAR instructs aircraft XYABC: "*XBC SQUAWK IDENT*". This means XBC shall operate (=*press*) the *IDENT-button* (Figure 3). The IDENT-button is on the transponder device at the cockpit panel. To be identified, the pilot should *press the button for approximately 2-3 seconds*.

 C) Wrong

If an identification turn should be performed, RADAR instructs the pilot: "*XBC perform identification turn of more than 20 degrees*".

 D) Wrong

When XBC shall *reselect his assigned mode and code*, the pilot is instructed for "*XBC recycle mode/code*".

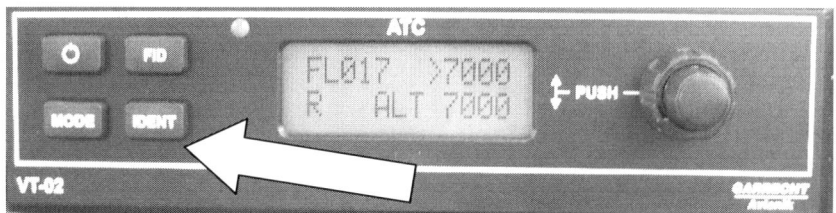

Figure 3: Transponder and IDENT-button

 Bei der Anweisung "Squawk ident" muss man den "Ident-Knopf" am Transponder zirka zwei bis drei Sekunden lang drücken.

120. RADAR instructs aircraft XYABC: "XBC SQUAWK STANDBY". What does this mean?

A)   XBC is requested to select the standby feature on the transponder

B)   XBC is requested to standby on the frequency

C)   XBC is requested to standby for radar vectors

D)   XBC is requested to standby as the radar controller is busy

A) Correct

If RADAR instructs aircraft XYABC "*XBC SQUAWK STANDBY*", the pilot should *switch the transponder-button to "standby"* (STBY or SBY, Figure 4). This stops a transponder submission, but facilitates rapid re-functioning of the device.

 B) Wrong

If the pilot is requested to *standby on the frequency*, ATC will instruct the pilot "XBC standby".

 C) Wrong

If the pilot is requested to *standby for radar vectors*, ATC will instruct the pilot "XBC standby (for radar vectors)".

 D) Wrong

If the pilot is requested to *standby as the radar controller is busy*, ATC will instruct the pilot "XBC standby".

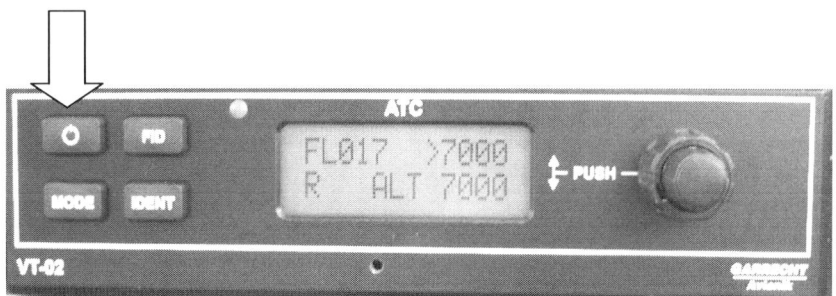

Figure 4: Standby (SBY) selector (sometimes in combination with on/off selector).

 Bei der Aufforderung "… squawk standby", muss man den Transponderknopf auf SBY (STBY) rasten.

121. RADAR instructs aircraft XYABC: "X-BC RESET SQUAWK 1015".
What does this mean?

A)   XBC is requested to set new code 1015

B)   XBC has been identified by SSR code 1015

C)   XBC is requested to reselect SSR code 1015

D)   XBC has been identified at 10:15

A) Wrong

If XBC is *requested to set new code 1015*, the pilot is instructed as "XBC squawk 1015".

 B) Wrong

If XBC has been *identified by SSR code 1015*, the controller will transmit "XBC identified".

 C) Correct

If XBC is *requested to reset squawk 1015*, the pilot must reselect SSR code 1015 and transmit this code again.

 D) Wrong

If XBC has been *identified at 10:15*, the controller will transmit "XBC identified at 1015".

 Bei der Meldung "… reset squawk …" muss man erneut den Transpondercode 1015 rasten.

Prüfungsvorbereitung für die Privatpilotenlizenz
Band 8B: Allgemein gültiges Sprechfunkzeugnis (AZF)
2. Auflage 2009

122. A pilot of an IFR flight has been instructed to establish radio contact with another ATC unit during climb. On initial contact he has to transmit the following data:

A) Call sign, present altitude/flight level and cleared altitude/flight level

B) Call sign only

C) Call sign and present altitude/flight level

D) Call sign and estimated time over the next compulsory reporting point

A) Correct

During *climb flight*, the pilot has to transmit the following details to ATC concerning his flight:

- Call sign,
- present altitude/flight level and
- cleared altitude/flight level

Example:

Pilot: D-EZOF passing altitude 4,000 ft climbing FL 120

 B) Wrong

Transmitting the *call sign only is not sufficient*, since present altitude/ flight level and cleared altitude/flight level have to be reported also.

 C) Wrong

Transmitting the call sign and present altitude/FL only is not sufficient, since the *cleared altitude/flight level has to be reported also.*

 D) Wrong

The *estimated time over the next compulsory reporting point* has not to be transmitted in the given situation.

 Beim Erstkontakt mit ATC im Steigflug müssen das Luftfahrzeug-kennzeichen, sowie die aktuelle bzw. freigegebene Flughöhe genannt werden.

123. To establish radio contact with "MÜNCHEN GROUND" the pilot of DIBEL shall trasmit the following call:

A)   MÜNCHEN GROUND DIBEL

B)   MÜNCHEN GROUND THIS IS DIBEL

C)   DIBEL, MÜNCHEN GROUND OVER

D)   MÜNCHEN GROUND DIBEL GO AHEAD

A) Correct

On initial contact with another station, *both the called and own sign have to be used*. In this situation, the called station has to be named first. Example:

Pilot: Mannheim GROUND D-EZOF or

Mannheim TOWER D-ECTE

 **B) Wrong**

The phrase "*this is*" must not be used.

 **C) Wrong**

The phrase "*over*" should not be used. Additionally, the *called station has to be named first*.

 **D) Wrong**

The phrase "*go ahead*" should not be used by the pilot.

 Wenn eine andere Station gerufen wird, muss zuerst der Name und die Bezeichnung der gerufenen und dann die eigene Funkstelle genannt werden.

124. A radio station of the aeronautical mobile service may simultaneously call discrete radio stations. This call is named:

A)   General call

B)   Urgency call

C)   simultaneous call

D)   Multiple call

A) Wrong

A "*general call*" is not defined for aviation communication.

 B) Wrong

A pilot has to transmit an "*urgency call*", if the plane or the occupants may be endangered and assistance is required.

 C) Wrong

A "*simultaneous call*" is not defined for aviation communication.

 D) Correct

A radio station of the aeronautical mobile service may simultaneously *call discrete radio stations*. This call is named "*multiple call*".

Example:

GROUND: D-ECTE, D-EZOF, D-EEDL taxi to Runway 27 via taxiways A and F.

 Bei einem "Meharfachanruf" werden Informationen gleichzeitig an mehrere Stationen (z.B. Luftfahrzeuge) übermittelt.

125. During approach to an airport with parallel runway system the pilot of an IFR flight has to transmit on initial contact, after changing frequency from approach control to aerodrome control, the radio callsign of his aircraft and...

A)   the designation of the runway being approached

B)   the type of instrument approach

C)   passing the transition level

D)   cleared altitude/flight level

A) Correct

It is essential to transmit in the presented situation the *designation of the runway being approached* to prevent *misunderstandings or confusion*. This applies only to approaches with two different parallel runways (e.g. 25L and 25R).

 B) Wrong

It is not required to transmit the *type of instrument approach*, since this information is less important as the designation of the runway.

 C) Wrong

It is not required to transmit *passing the transition level*, since this information is less important as the designation of the runway.

 D) Wrong

Cleared altitude/flight level has to be confirmed during climb flight only. Since the situation describes an approach, this information should not be transmitted.

 Beim Erstkontakt mit TOWER muss das Kennzeichen und die Bezeichnung der angeflogenen Piste genannt werden.

126. The pilot of an IFR flight has to transmit the following data on initial contact after changing frequency from approach control to aerodrome control:

A)   Radio callsign of the aircraft only

B)   Aircraft call sign and flight level or altitude

C)   Aircraft call sign and cleared altitude/flight level

D)   Aircraft call sign and present position

A) Correct

When changing *frequency from approach control to aerodrome control*, the pilot has to transmit on initial contact his *radio callsign of the aircraft*. If two different parallel runways are in use, the designation of the runway has to be transmitted additionally.

 B) Wrong

The *flight level or altitude* has to be confirmed during climb flight only. Since the situation describes an approach, this information should beommitted.

 C) Wrong

Cleared altitude/flight level has to be confirmed during climb flight only. Since the situation describes an approach, this information should not be transmitted.

 D) Wrong

The *present position* should not be reported, since this may be gathered on the radar screen by the controller.

 Beim Anflug und Erstkontakt mit TOWER muss lediglich das Rufzeichen des Luftfahrzeugs genannt werden.

127. When establishing communication, how shall aircraft XYABC call Stephenville TOWER?

A)   Stephenville TOWER XBC

B)   Stephenville TOWER XYABC

C)   Stephenville XYABC

D)   TOWER XYABC

A) Wrong

For the initial contact, the *full aircraft radio callsign* has to be used!

 B) Correct

When establishing communication, the *name, designation, and the complete own radio callsign* has to be transmitted. *Example*:

Pilot: Mannheim TOWER D-EZOF

TOWER: D-EZOF Mannheim TOWER

 C) Wrong

The *designation* of the called station (i.e. TOWER) has to be called also.

 D) Wrong

The *name* (*callsign*) of the station (i.e. TOWER) has to be called also.

 Beim Erstkontakt muss der Name, die Bezeichnung und das vollständige eigene Rufzeichen genannt werden.

128. Aircraft XYABC has been instructed to contact Stephenville TOWER on frequency 118,7. What is the correct way to indicate it will follow this instruction?

A)   Will change to TOWER XBC

B)   XBC 118.7

C)   XBC changing over

D)   Stephenville TOWER XBC

A) Wrong

The word "*to*" should not be used for communication since it *may be mistaken by "two"*.

 B) Correct

From all given answers *this is the one best suited*. From the question it is not clear, if the *abbreviated callsign* (X-BC) may be used. Additionally the phrase "*...will change frequency...*" might be used. Therefore, another correct answer would be:

Pilot: X-YABC will change frequency 118.7

 C) Wrong

"*Changing over*" is not an official phrase for communication.

 D) Wrong

"*Stephenville TOWER XBC*" is the correct phrase for an initial contact *but not to confirm frequency change*.

 Beim Frequenzwechsel genügt es, das eigene Rufzeichen in Verbindung mit der neuen Frequenz zu nennen.

129. Aircraft XY-ABC has been instructed to listen on ATIS frequency 123.250, on which the aerodrome data are being broadcast. What is the correct way to indicate it will follow this instruction?

A)   XBC monitoring 123.250

B)   Changing to 123.250 XBC

C)   Will contact 123.250 XBC

D)   XBC checking 123.250

A) Correct

To indicate, a pilot will *follow the given instruction*, he shall use the phrase (*MONITOR* = "I will follow the given instructions"). In This case, the pilot has to answer: *X-BC MONITORING 123.250*. Also right is the answer X-BC will monitor 123.250.

Beeing critical, the provided answer is not completely correct. XBC may only be used, *if previously used by ATC*. This fact is not reported in the question/answer. Additionally, the *word "frequency"* should be used. Therefore we recommend using the following phrase:

XY-ABC MONITORING frequency 123.250.

 B) Wrong

"*Changing*" is not an existing phrase. Additionally, the word "*to*" should be omitted whenever possible due to confusion with the word "*two*".

 C) Wrong

The pilot was asked to monitor a frequency, but *not to contact someone*.

 D) Wrong

"*Checking*" is not an existing phrase.

 Die Sprechgruppe MONITOR besagt, dass man eine Frequenz abhören soll.

130. Aircraft **XYABC** is making a test transmisstion with Stephenville **TOWER** on frequency 118.7. What is the correct phrasing for this transmission?

A)   Stephenville TOWER XYABC preflight check

B)   Stephenville TOWER XYABC signal check

C)   Stephenville TOWER XYABC how do you read

D)   Stephenville TOWER XYABC frequency check

A) Wrong

A *preflight check* has to be performed manually before starting the engines.

 B) Wrong

A *signal check* is not defined for aviation.

 C) Correct

The correct phrasing for a *test transmission* on a frequency is *"how do you read?"*. This indicates, a pilot *verifies proper radio communication* with the tower.

 D) Wrong

A *frequency check* is not defined for aviation.

 Die Sprechgruppe zum prüfen der Funktionsfähigkeit des Funks lautet "how do you read?".

131. On the readability scale what does "readability 3" mean?

A)   No problem to understand

B)   Loud and clear

C)   Readable but with difficulty

D)   Unreadable

A) Wrong

If communication is *no problem*, "*readability 5*" has to be used.

 B) Wrong

If communication is no problem (i.e. *loud and clear*), "*readability 5*" has to be used.

 C) Correct

"*Readability 3*" indicates *communication is possible but with difficulties* (Table 5).

 D) Wrong

If communication is *unreadable*, it is "readability 1".

Readability	Meaning
1	Unreadable
2	Nearly not readable
3	Readable but with difficulty
4	Readable
5	No problem to understand, "loud and clear", perfectly readable

Table 5: Readability scale for radio communication

132. On the readability scale what does "readability 5" mean?

A)   Unreadable

B)   Problem to understand

C)   Perfectly readable

D)   Readable but with difficulty

A) Wrong

If communication is *unreadable*, it is "readability 1".

 B) Wrong

If communication is a "*problem to understand*", it has to be classified as "nearly not readable", i.e. "readability 2".

 C) Correct

If communication is *no problem (i.e. perfectly readable)*, "*readability 5*" has to be used (Table 5, page 271).

 D) Wrong

"*Readability 3*" indicates *communication is possible but with difficulties.*

133. A compulsory reporting point is a defined location where a position report must be made...

A)   on request of ATC

B)   in VMC only

C)   in IMC only

D)   in any case

A) Wrong

A defined location where a position report must be made *on request of ATC* is a *"non-compulsory* reporting point".

 B) Wrong

Reporting points do not exist for *VMC* (VFR) traffic only.

 C) Wrong

Reporting points do not exist for *IMC* (IFR) traffic only.

 D) Correct

Position reports *must be made in any case* when overflying a *compulsory reporting point* (Figure 5). This accounts for both VFR and IFR flights.

Figure 5: Symbol for a compulsory reporting point.

134. A non-compulsory reporting point is a defined location where a position report must be made...

A)   on request of ATC

B)   in any case

C)   in VMC

D)   in IMC

A) Correct

A *non-compulsory reporting point* is a defined location where a *position report must be made on request of ATC only* (Figure 6).

 B) Wrong

Position reports *must be made in any case* when overflying a *compulsory reporting point* (Figure 5, page 275). This accounts for both VFR and IFR flights.

 C) Wrong

Reporting points do not exist for *VMC* (VFR) traffic only.

 D) Wrong

Reporting points do not exist for *IMC* (IFR) traffic only.

Figure 6: Symbol for a non-compulsory reporting point.

135. **Which elements of information should an abbreviated position report during an IFR flight always contain?**

A)   Aircraft identification, position, time, level

B)   Aircraft identification, position, time

C)   Aircraft identification, position, level

D)   Aircraft identification, position, next position

A) Wrong

Reporting the *altitude/flight level* is not necessary in an abbreviated position report, since both are encoded by the transponder.

 B) Correct

The following elements should be reported in an abbreviated position report:

- *Aircraft identification,*
- *position, and*
- *time*

 C) Wrong

Reporting the *altitude/flight level* is not necessary in an abbreviated position report, since both are encoded by the transponder.

 D) Wrong

Reporting the *next position* is not necessary in an abbreviated position report.

 Abgekürzte Positionsmeldungen bei einem IFR-Flug sollten die folgenden Informationen enthalten: Luftfahrzeugkennung, Position und Zeit.

136. What shall be the pilot's readback for: "CLIMB FL 280"?

A)   Climbing to flight level two eight zero

B)   Climbing two eight zero

C)   Climbing to two eighty

D)   Climbing flight level two eight zero

A) Wrong

It is not authorized to say "… *to* …", since it may be mistaken as "… *two* …" and may therefore cause problems.

 B) Wrong

The phrase "*Flight Level*" has to be read back.

 C) Wrong

Numbers have to be spoken as separate *digits*. Additionally, it is not authorized to say "… *to* …", since it may be mistaken as "… *two* …" and may therefore cause problems.

 D) Correct

The correct readback is: "*Climbing flight level two eight zero*". Instead, "Will climb flight level two eight zero" is also possible.

 Das Wort "to" sollte wegen der Verwechselungsgefahr mit "two" nicht verwendet werden!

137. What shall the pilot's readback be for: "CLIMB ALTITUDE 2,500 FEET"?

A)   Up two thousand five hundred

B)   Climbing to two point five

C)   Climbing altitude two thousand five hundred feet

D)   Climbing to altitude two thousand five hundred feet

A) Wrong

The phrase "*climb*" should be read back.

 B) Wrong

The phrase "*to two point five*" can be misinterpreted as 22.5 and shall not be used.

 C) Correct

The correct readback is "*Climbing altitude two thousand five hundred feet*". Alternatively, the readback "Will climb altitude two thousand five hundred feet" is acceptable.

 D) Wrong

It is not authorized to say "... *to* ...", since it may be mistaken as "... *two* ..." and may therefore cause problems.

138. ATC clears Fastair 345 to descend from FL 100 to FL 80. What is the correct readback by the pilot?

A)   Descending to 80, Fastair 345

B)   Down to flight level 80, Fastair 345

C)   Fastair 345 leaving 100 to 80

D)   Fastair 345 leaving flight level 100 descending flight level 80

A) Wrong

The phrase "*Flight Level*" has to be read back. Additionally, the present flight level (FL 100) has to be reported.

 B) Wrong

The *present flight level* (FL 100) has to be reported. Additionally, the word "down" should not be used.

 C) Wrong

The phrase "*Flight Level*" has to be read back.

 D) Correct

The correct readback is "*Fastair 345 leaving flight level 100 descending flight level 80*".

 "Flight level" oder "Altitude" muss immer zurückgelesen werden!

Prüfungsvorbereitung für die Privatpilotenlizenz
Band 8B: Allgemein gültiges Sprechfunkzeugnis (AZF)
2. Auflage 2009

139. Runway visual range (RVR) is included in the weather report when the visibility is...

A)   1,000 ft or less

B)   less than 1,500 m

C)   less than 1,000 m

D)   1,500 ft or less

 **A) Wrong**

The *RVR* is reported if *visibility is less than 1,500 m* (not 1,000 ft).

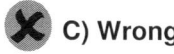 **B) Correct**

The *RVR* is reported if *visibility is less than 1,500 m.*

 C) Wrong

The *RVR* is reported *if visibility is less than 1,500 m* (not 1,000 m).

D) Wrong

The *RVR* is reported if *visibility is less than 1,500 m* (not 1,500 ft).

 Die RVR wird im Wetterbereicht separate genannt, wenn die Sicht unterhalb 1.500 m liegt.

140. A pilot of an IFR flight shall inform the ATC unit competent for approaches and departures that he has received ATIS...

A)   if deemed necessary

B)   on request only

C)   if the control zone is IMC

D)   upon initial contact

A) Wrong

Informing ATC about a received ATIS report is a mandatory part of the initial call and is not performed *if deemed necessary.*

 B) Wrong

A pilot has to inform ATC automatically (*not on request*) about a received ATIS upon initial contact.

 C) Wrong

A pilot has to inform ATC automatically about a received ATIS upon initial contact (*and not only if the control zone is IMC*).

 D) Correct

A pilot has to *inform ATC automatically* about a received ATIS *upon initial contact.*

 Ein Pilot muss ATC beim ersten Kontakt über den Erhalt der aktuellen ATIS-Medlung informieren.

141. When the term "scattered" is used in radio telephony in connection with meteorological conditions, the amount of clouds covering the sky is...

A)   more than half but less than overcast (5 to 7 octas)

B)   sky entirely covered (8 octas)

C)   no clouds below 5,000 feet/GND

D)   half or less than half (3 to 4 octas)

A) Wrong

If the sky is covered by clouds *with more than half but less than overcast* (5 to 7 octas), it is defined as "broken" (BKN).

 B) Wrong

If the sky is *entirely covered by clouds* (8/8 or 100 %) it is rated "overcast" (OVC).

 C) Wrong

It does not matter in *which altitude clouds are present.* If there are no clouds, it is rated "*sky clear*" (SKC).

 D) Correct

If the sky is *half or less than half* (3/8 to 4/8) covered by clouds, it is defined as "*scattered*" (SCT; Table 6).

Clouds	Means	Abbreviation
0/8	Sky clear	SKC
1/8 to 2/8	Few	FEW
3/8 to 4/8	Scattered	SCT
5/8 to 7/8	Broken	BKN
8/8	Overcast	OVC
9/8	Obscured	OBSC

Table 6: Definition of cloud layers and abbreviations

142. When the term "broken" is used in radio telephony in connection with meteorological conditions, the amount of clouds covering the sky is:

A)   1 to 4 octas

B)   8 octas below 10,000 feet

C)   5 to 7 octas

D)   No clouds below 5,000 feet

A) Wrong

1/8 to 4/8 clouds in the sky may be classified with *"few"* (FEW) or *"scattered"* (SCT).

 B) Wrong

If the *sky is covered by 8 octas* (8/8; idenpendent from altitude), it is classified as *"overcast"* (OVC).

 C) Correct

When using the term *"broken"* (BKN), *5/8 to 7/8* of the sky are covered by clouds.

 D) Wrong

It does not matter in *which altitude clouds are present.* If there are no clouds, it is rated *"sky clear"* (SKC).

 Bei einem Bedeckungsgrad von "broken" (BKN) sind 5/8 bis 7/8 des Himmels mit Wolken bedeckt!

143. When the term "overcast" is used in radio telephony in connection with meteorological conditions, the amount of clouds covering the sky is:

A)   100 %

B)   50 % or more

C)   Less than 50 %

D)   No clouds but poor ground visibility

A) Correct

The term "*overcast*" (OVC; Table 6, page 291) means that the whole sky (*8/8 or 100 %*) is covered with clouds.

 B) Wrong

If the sky is covered by 50% or more (4/8 or more), the terms "*scattered*" (SCT) or "*broken*" (BRK) are used.

 C) Wrong

If the sky is covered by less than 50% (i.e. less than 4/8), the terms "*scattered*" (SCT) or "*few*" (FEW) are used.

 D) Wrong

If no clouds are visible, the term "*sky clear*" (SKC) is used. With this definition, ground visibility cannot be classified.

 Beim Bedeckungsgrad "overcast" (OVC) ist der komplette Himmel (100 % oder 8/8) mit Wolken bedeckt!

144. When the term "CAVOK" is used in an aviation routine weather report (METAR), the values of visibility and clouds are:

A)   Visibility 10 km or more, no clouds below 5,000 feet/GND

B)   Visibility 10 km or more, no clouds below 1,500 feet/GND

C)   Visibility more than 8 km, no clouds below 3,000 feet/GND

D)   Visibility more than 5,000 m, no clouds below 1,500 m/GND

A) Correct

The term "CAVOK" means *"Clouds and visibility o.k."* or *"ceiling and visibility o.k."*. In detail, CAVOK conditions exist if *ground visibility* is > 10 km, *clouds* are not below 5,000 ft and *precipitation*s as well as thunderstorms are absent.

 B) Wrong

Clouds have to be above 5,000 ft not 1,500 ft using the CAVOK definition.

 C) Wrong

Visibility has to be more than 10 km (not 8 km) and clouds have to be above 5,000 ft (not 3,000 ft) when using the CAVOK definition.

 D) Wrong

Visibility has to be more than 10 km (not 5,000 m) and clouds have to be above 5,000 ft (not 1,500 m) when using the CAVOK definition.

145. In what units of measurement is the visibility in an aviation routine weather report (METAR) expressed in plain language?

A)   Up to 5,000 m in metres, above in kilometres

B)   Up to 1,500 m in metres, above in kilometres

C)   In feet and nautical miles

D)   In nautical miles only

A) Correct

Horizontal visibility of less than 5,000 m has to be expressed in plain language in *meters (m),* e.g. in a METAR. A visibility above 5,000 m is given in *kilometers (km).*

 B) Wrong

The *cut-off point* for using meters (m) is 5,000 m (not 1,500 m).

 C) Wrong

Feet (ft) and *nautical miles* (NM) may be used for vertical/horizontal distances between two geographical points, but not for visibility.

 D) Wrong

Nautical miles (NM) may be used for horizontal distances between two geographical points, but not for visibility.

 Bei der Sichtweite liegt die Grenze zur Angabe in Meter oder Kilometer bei 5.000 m!

146. What is the correct way of expressing visibility in plain language (in METAR)?

A)   Visibility 1,200 feet

B)   Visibility 1.2 nautical miles

C)   Visibility 1.2 kilometres

D)   Visibility 1,200 metres

A) Wrong

Horizontal visibility must not be expressed in feet, but in meters (m) or kilometers (km). Therefore, 1,200 ft is incorrect.

 B) Wrong

Horizontal visibility must not be expressed in nautical miles (NM), but in meters (m) or kilometers (km). Therefore, 1.2 NM is incorrect.

 C) Wrong

Horizontal visibility of less than 5 km has to be expressed in meters (m). Therefore, *1.2 km* is incorrect.

 D) Correct

Horizontal visibility of less than 5,000 m has to be expressed in plain language in *meters (m),* e.g. in a METAR. A visibility above 5,000 m is given in *kilometers (km).* Since 1,200 m is less than 5 km, the answer is correct.

 Eine Horizontalsicht wird unterhalb von 5.000 m in Metern (m) angegeben, darüber in Kilometern (km).

147. When transmitting runway visual range (RVR) for runway 16, ATC will use the following phrase:

A)   The values of the transmissiometer are: ... metres and ... metres

B)   RVR runway 16 touchdown zone ... metres, mid-point ... metres, stop end ... metres

C)   RVR at the beginning of runway 16 is ... metres

D)   RVR runway 16 ... metres diagonal ... metres diagonal ... metres

A) Wrong

The RVR is given *at three but not two defined points* for the runway currently in use.

 B) Correct

The *RVR* (Runway Visual Range) is the *horizontal visibility direct on the runway*. It is given *three times*, i.e. at the *touchdown* zone, *mid*-point, and at the *stop end*. RVR is given in meters (below 5,000 m) or in kilometres (above 5 km).

Example: D-EZOF, RVR runway 16 touchdown zone 800 metres, mid-point 1,200 metres, stop end 1,800 metres.

 C) Wrong

The RVR is given *at three, not one defined points* for the runway currently in use.

 D) Wrong

The RVR is given *at three but not two defined points* for the runway currently in use.

 Die Runway Visual Range (RVR) ist die horizointale Landebahn-sicht. Sie wird an drei definierten Punkten erfasst und an den Luftfahrzeugführer übermittelt.

148. What does "friction coefficient 45" in a runway report mean?

A)   Braking action medium

B)   Braking action good

C)   Braking action poor

D)   Braking action not measurable

A) Wrong

A *friction coefficient 45* is good but not medium (30 to 35).

 B) Correct

The friction coefficient (Table 7) is a *dimensionless quantity used to calculate the force of friction* (*static* or kinetic). It is measured with the Tapley-Meter, Skidometer, the Saab-Friction-Tester and the μ-Meter on each runway side. The coefficient of *static* friction is defined as ratio of the *maximum static friction* force (F) between the surfaces in contact to the *normal* (N) force.

A friction coefficient 45 indicates *good braking* action.

 C) Wrong

A friction coefficient 20 - but not 45 - indicates *poor braking* action.

 D) Wrong

If the braking action is *not measureable*, it is reported as "not measureable" or "unreliable".

Friction coefficient	Description
> 40	Good
36 to 39	Medium to good
30 to 35	Medium
26 to 29	Poor to medium
<25	poor
Unreliable	unreliable

Table 7: Friction coefficients

149. What does "friction coefficient 20" in a runway report mean?

A)   Braking action medium

B)   Braking action good

C)   Braking action poor

D)   Braking action unreliable

A) Wrong

A *friction coefficient 20* is poor but not medium (30 to 35).

 B) Wrong

Good braking action is equal to a high friction coefficient (> 40).

 C) Correct

The coefficient of friction (Table 7, page 305) is a *dimensionless quantity used to calculate the force of friction* (*static* or kinetic). It is measured with the Tapley-Meter, Skidometer, the Saab-Friction-Tester and the µ-Meter on each runway side. The coefficient of *static* friction is defined as the ratio of the *maximum static friction* force (F) between the surfaces in contact to the *normal* (N) force. A friction coefficient 20 indicates *poor braking* action.

 D) Wrong

Unreliable braking action is equal to an unreliable friction coefficient.

150. Under which runway conditions is the braking action reported to be "unreliable"?

    A)   Runway covered with wet snow and slush

    B)   Runway covered with ice

    C)   Runway covered with dry snow

    D)   Runway conditions normal

A) Correct

If the runway is covered with *snow and slush*, the friction coefficient is always unreliable (not accurate measurement possible).

 B) Wrong

A friction coefficient is reported, if the runway is *covered with ice*.

 C) Wrong

A friction coefficient is reported, if the runway is *covered with dry snow*.

 D) Wrong

For *normal runway conditions*, the friction coefficient is given as a number.

 Bei einem Bremskoeffizient < 30 muss mit schlechten Bremsverhältnissen gerechnet werden.

151. If you are requested to "REPORT FLIGHT CONDITIONS", what does that mean?

A)   Indicate whether you are flying in IMC or in VMC

B)   Indicate weather conditions as wind, visibility, temperature

C)   Indicate if visibility is sufficient for landing

D)   Indicate whether you are flying IFR or VFR

A) Correct

There are defined flight conditions: VMC (visual meteorological conditions) and IMC (Instrument meteorological conditions).

VMC means that all required minima for flights according to VFR are applicable within the present airspace; IMC means that these minima are not met.

 B) Wrong

Flight conditions (VFR, IFR) may not be confused with weather conditions (VMC/Visual meteorological conditions or IMC/Instrument meteorological conditions).

 C) Wrong

Flight conditions may not be confused with weather conditions (e.g. visibility).

 D) Wrong

There are two defined flight rules: VFR and IFR. Flights may be performed either according to visual flight rules (VFR) or instrument flight rules (IFR). Flight rules may not be confused with flight conditions.

VFR means that the pilot has to maintain separation by himself and must have visual control of the airspace. IFR means that the aircraft is totally controlled (e.g. by ATC) and the pilots performs the flight according to instructions given by the controller and the indication of his instruments.

152. An aeronautical station using the identification "VOLMET" in its callsign...

A)   can be called by an aircraft in flight to obtain flight information service

B)   executes air traffic control service to enroute aircraft

C)   is an aeronautical station operated by an airport company

D)   is a broadcasting service for the transmission of aerodrome weather reports and landing forecasts

A) Wrong

An aeronautical station using the identification *VOLMET* (Meteorological Information for Aircraft In Flight) transmits meteorological reports *automatically and continuously* and can therefore not answer.

 B) Wrong

ATC (Air Traffic Control) executes air traffic control service to enroute aircraft using the call sign *RADAR*.

 C) Wrong

Airport companies do not provide meteorological information for aircraft in flight (VOLMET).

 D) Correct

An aeronautical station using the identification "*VOLMET*" in its callsign is a *broadcasting service for the transmission of aerodrome weather reports* and landing forecasts (METAR). VOLMET reports contain actual weather conditions at different airports, e.g. METAR reports.

 Wettermeldungen (METAR) werden automatisch und kontinuierlich auf VOLMET-Frequenzen ausgestrahlt.

153. The weather report in an ATIS broadcast contains the term "CAVOK". This means that an arriving aircraft has to expect...

A)   less than 5/8 clouds below 5,000 ft

B)   light precipitation

C)   thunderstorm

D)   no clouds below 5,000 ft GND

A) Wrong

Instead of less than *5/8 clouds below 5,000 ft* (Table 6, page 291), CAVOK means no clouds below 5,000 ft!

 B) Wrong

When noticing "*light precipitation*" (e.g. rain), the term CAVOK may not be used.

 C) Wrong

When noticing a *thunderstorm*, the term CAVOK may not be used.

 D) Correct

The term "CAVOK" means "*Clouds and visibility o.k.*" or "*ceiling and visibility o.k.*". In detail, CAVOK conditions exist, if *ground visibility* is > 10 km, *clouds* are not below 5,000 ft and *precipitation* and *thunderstorms* are absent.

 CAVOK-Kriterien: Bodensicht > 10 km, keine Wolken unter 5.000 ft, keine signifikanten Wettererscheinungen (z.B. Niederschlag) und keine Gewitterwolken.

154. What is normally used for ATIS broadcast?

A)   VOR frequencies and/or discrete VHF frequencies

B)   Voice channel of an ILS

C)   NDB frequencies

D)   DME voice channels

A) Correct

ATIS (Automatic Terminal Information Service) reports are *transmitted continuously on VOR-frequncies or separate ATIS-frequencies* and contain relevant acutual weather information for take-off and landing.

Example: This is Mannheim Airport, Information Alpha, Met Report Time 1620, LOC-DME approach runway 27, transition level 60, wind 240 degrees, 5 knots, visibility 9 kilometers, light rain, clouds scattered 3,000 feet, temperature 24, dewpoint 18, QNH 1018, NOSIG, Information Alpha out.

 B) Wrong

ATIS reports are not broadcasted on *ILS* (Instrument Landing System) *frequencies*. Instead, a unique morse code for identification is broadcasted on these frequencies.

 C) Wrong

ATIS reports are not broadcasted on *NDB* (Non-Directional Beacon) *frequencies*. Instead, a unique morse code for identification is broadcasted on these frequencies.

 D) Wrong

ATIS reports are not broadcasted on *DME* (Distance Measuring Equipment) *frequencies*. Instead, a unique morse code for identification is broadcasted on these frequencies.

 ATIS-Medlungen werden normalerweise auf VOR-Frequenzen oder separaten ATIS-Frequenzen ausgetrahlt.

155. How can aviation routine weather reports (METAR) of specific airports be obtained by aircraft in flight?

A)   ATIS

B)   AFIS

C)   SIGMET

D)   VOLMET

A) Wrong

METAR (Meteorological Aviation Routine Weather Report) reports are *not transmitted on ATIS* (Automatic Terminal Information Service) *frequencies.* Instead, METAR reports have to be obtained on VOLMET (Meteorological Information for Aircraft In Flight) frequencies.

 B) Wrong

AFIS is *no abbreviation used* in aviation.

 C) Wrong

SIGMETs (significant meteorological phenomena) only provide *information on significant meteorological phenomena*, e.g. hail, thunderstorm, sand storms or icing.

 D) Correct

METAR (Meteorological Aviation Routine Weather Report) reports are transmitted on *VOLMET* (Meteorological Information for Aircraft In Flight) frequencies. METAR reports inform pilots about *relevant meteorological conditions* at different airports.

Example:
EDDF 081520Z 19010KT 9999 FEW040TCU 09/M03 Q1012 NOSIG

 METAR-Meldungen können auf den veröffentlichten VOLMET-Frequenzen empfangen werden.

156. Which information can aircraft in flight obtain by VOLMET?

A)   SPECI and TAF

B)   SIGMET

C)   Aviation routine weather reports (METAR) of specific airports

D)   Runway reports

A) Wrong

TAF (Terminal Aerodrome Forecast) *reports* are not transmitted on VOLMET frequencies.

 B) Wrong

SIGMETs (significant meteorological phenomena) only provide *information on significant meteorological phenomena*, e.g. hail, thunderstorm, sand storms or icing.

 C) Correct

METAR (Meteorological Aviation Routine Weather Report) reports are transmitted on *VOLMET* (Meteorological Information for Aircraft In Flight) frequencies. METAR reports inform pilots about *relevant meteorological conditions* at different airports.

 D) Wrong

Runway reports are not transmitted on VOLMET frequencies.

157. Distress is defined as...

A)   a condition concerning the safety of an aircraft or of a person on board, but which does not require immediate assistance

B)   a condition concerning the attitude of an aircraft when intercepting the localizer during an ILS approach

C)   a condition of being threatened by serious and/or imminent danger and of requiring immediate assistance

D)   a condition concerning the safety of a person on board or within sight and requiring immediate assistance

A) Wrong

An urgency situation is defined as a condition concerning the safety of an aircraft or of a person on board, but which does not require immediate assistance

 B) Wrong

Distress has nothing to do with an aircraft *intercepting the localizer during an ILS approach.*

 C) Correct

Distress is defined as a *condition of being threatened by serious and/or imminent danger and of requiring immediate assistance.* Examples for distress situations are fire onboard, explosions onboard, an emergency landing or severe engine trouble. A distress situation endangers both airplane and occupants.

 D) Wrong

An urgency situation is defined as a condition concerning the safety of a person on board or within sight and requiring immediate assistance.

 Bei einer Notlage sin dimmer das Luftfahrzeug und die Insassen gefährdet, nicht nur eine einzelne Person!

158. A signal sent by radiotelephony consisting of the spoken word MAYDAY means:

A)   The aircraft has a very urgent message to transmit concerning the safety of a ship, aircraft or vehicle

B)   The aircraft has a message to transmit concerning adverse weather conditions along its route or flight

C)   Imminent danger threatens the aircraft and immediate assistance is required

D)   The aircraft is forced to perform a fuel dumping procedure

A) Wrong

If the aircraft has a *very urgent message* to transmit, the pilot shall use the phrase *PAN PAN*, preferably spoken three times at the beginning of the message.

 B) Wrong

Adverse weather reports may be transmitted as *normal messages* or as *urgency message*, e.g. if a VFR flight has to be conducted during IFR conditions.

 C) Correct

If an aircraft is *endangered* by an *emergency*, the pilot has to transmit an *emergency (distress) message.* This is indicated by the word *MAYDAY*, preferably spoken three times at the beginning of the message. In such a message the pilot shall report the problem, altitude, position and further actions being taken.

Example: MAYDAY MAYDAY MAYDAY D-EZOF engine out, performing emergency landing, altitude 3,000 ft, position 5 NM north of Mannheim airfield.

 D) Wrong

A *fuel dumping procedure* is not necessarily an emergency.

 Eine Notmedlung wird mit der Sprechgruppe MAYDAY (am besten dreimal hintereinander) eingeleitet.

159. An aircraft in distress shall send the following signal by radiotelephony:

A)   DETRESFA, spoken three times

B)   MAYDAY, preferable spoken three times

C)   PAN PAN, spoken three times

D)   URGENCY, spoken three times

A) Wrong

DETRESFA is the distress phase being established by the Search And Rescue (SAR) Service. It is not a *signal for radiotelephony*.

 B) Correct

If an aircraft is *endangered* by an *emergency*, the pilot has to transmit an *emergency* (*distress*) *message*. This is indicated by the word *MAYDAY*, preferable spoken three times at the beginning of the message. In such a message the pilot shall report the problem, altitude, position and further actions being taken.

 C) Wrong

If the aircraft has a *very urgent message* to transmit, the pilot shall use the phrase *PAN PAN*, preferably spoken three times at the beginning of the message.

 D) Wrong

The word URGENCY is not used for radiotelephony.

160. The distress signal and the distress message to be sent by an aircraft in distress shall be on...

A)   the air-ground frequency in use at the time

B)   the emergency frequency in any case

C)   the regional guard frequency

D)   the FIS frequency designated for the airspace concerned

A) Correct

A message starting with the phrase *"MAYDAY"* is an emergency message or *distress message*. Emergency messages should be transmitted *on the active frequency*. Changing to another frequency may cost valuable time in an emergency.

 B) Wrong

Emergency messages should be transmitted on the *active frequency*, but may be transmitted on the *international emergency frequency* (121,500 MHz) also. Changing frequency prior to transmission is not recommended.

 C) Wrong

Emergency messages should be transmitted on the *active frequency*. Changing frequency to the *regional guard frequency* prior to transmission is not recommended.

 D) Wrong

Emergency messages should be transmitted on the *active frequency*. Changing frequency to the *FIS (Flight Information Service) frequency* prior to transmission is not recommended.

161. The frequency used for the first transmission of a "MAYDAY" call shall be...

A)   in use at that time or on an emergency frequency

B)   the distress frequency 121.5 MHz

C)   any other international emergency frequency

D)   any frequency at pilot's discretion

A) Correct

A message starting with the phrase "*MAYDAY*" is an emergency message or *distress message*. Emergency messages should be transmitted *on the active frequency or on the international emergency frequency (121.500 MHz)*. The word "MAYDAY" should be transmitted three times before the message.

 B) Wrong

Emergency messages should be transmitted on the active frequency, but may be transmitted on the *international emergency frequency* (121.500 MHz) also. Changing frequency prior to transmission is not necessary.

 C) Wrong

Emergency messages should be transmitted as soon as possible. Therefore, *changing frequency is not required*.

 D) Wrong

Emergency messages should be transmitted *on the active frequency*, currently in use, and not on a frequency at pilot´s discretion.

 Notmedlungen sollten so rasch als möglich auf der gerade aktiven Frequenz oder der internationalen Notfrequenz übermittelt werden!

162. The distress message shall contain at least the following elements / details:

A) Aircraft call sign, route of flight, destination airport

B) Aircraft call sign, nature of distress, pilot's intention, present position, level and heading

C) Aircraft call sign, aerodrome of departure, position and level

D) Aircraft call sign, present position, assistance required

A) Wrong

With an emergency or distress message important information about the emergency should be transmitted as short and precise as possible. Therefore, *route of flight* and *destination airport* have to be omitted.

 B) Correct

With an emergency or distress message important information about the emergency should be transmitted *as short and precise as possible.* Therefore, the following details are essential:

- *aircraft call sign,*
- *nature of distress,*
- *pilot's intention, as well as*
- *present position, level and heading*

 C) Wrong

With an emergency or distress message important information about the emergency should be transmitted as short and precise as possible. Therefore, the *aerodrome of departure* has to be omitted.

 D) Wrong

All the given details are essential for an emergency message, but some other *information is missing.*

 Eine Notmedlung muss das Rufzeichen des Luftfahrzeugs, die Art der Notlage, die Absicht des Piloten, sowie die aktuelle Position, Höhe und Kurs enthalten!

163. Which of the following frequencies is an international emergency frequency?

A)   122.500 MHz

B)   121.500 MHz

C)   6,500 KHz

D)   121.050 MHz

A) Wrong

122.500 MHz is not an international emergency frequency.

 B) Correct

For civil aviation, the *emergency frequency* is internationally defined as *121.500 MHz*. For *military aviation*, the emergency frequency is *243.000 MHz* (Table 8, page 337).

 C) Wrong

6,500 kHz is not an international emergency frequency.

 D) Wrong

121.050 MHz is not an international emergency frequency.

 Neben der zivilen Notfrequenz gibt es auch eine militärische!

164. The frequency 121.500 MHz is...

A)   an international emergency frequency

B)   a frequency for air-to-air communication

C)   a regional VHF emergency frequency

D)   a regional guard frequency

A) Correct

For civil aviation, the *emergency frequency* is internationally defined as *121,500 MHz*. For *military aviation*, the emergency frequency is *243,000 MHz* (Table 8).

 B) Wrong

For *air-to-air communication*, a pilot may use the freuquency *122,800 MHz* in Germany (AIP-VFR GEN 1-15).

 C) Wrong

Regional VHF (Very High Frequency) *emergency frequencies* are not published.

 D) Wrong

121,500 MHz is the international emergency frequency, not a *regional guard frequency*.

Type	Civil aviation	Military aviation
Emergency frequency	121,500 MHz	243,000 MHz
Air-to-air frequency	122,800 MHz	

Table 8: Frequencies for civil and military aviation.

 Die internationale zivile Notfrequenz lautet 121,500 MHz, für die Militärluftfahrt 243,000 MHz.

165. An aircraft in a distress situation shall squawk:

A)   7700

B)   6700

C)   7600

D)   7500

A) Correct

An aircraft in *distress* (e.g. complete engine failure, fire on board etc) shall squawk the internationally defined *mode/code A/C 7700* (Table 9, page 357).

 B) Wrong

The code *6700* is not specifically defined for Germany.

 C) Wrong

The transponder code *7600* indicates *radio communication failure* not a distress situation.

 D) Wrong

The transponder code *7500* indicates *hijacking* not a distress situation.

 Bei einer Notlage muss der Transpondermode/-code A/C/S7700 gerastet werden!

166. An aircraft squawking 7700 indicates to the ground station that...

A)   the aircraft is in distress

B)   the aircraft is being hijacked

C)   the aircraft's transceiver is unserviceable

D)   there is a very sick passenger on board

A) Correct

An aircraft in *distress* (e.g. complete engine failure, fire on board etc) shall squawk the internationally defined *mode/code A/C 7700* (Table 9, page 357).

 B) Wrong

The transponder code *7500* indicates *hijacking.*

 C) Wrong

The transponder code *7600* indicates *radio communication failure.*

 D) Wrong

Having a *very sick passenger* on board, the pilot may send an *urgency message* (PAN PAN). Since the aircraft itself is *not endangered*, this is not a distress situation.

167. Under which of the following circumstances shall an aircraft squawk an internationally prescribed mode/code?

A)   when following a SID

B)   when flying within controlled airspace

C)   when passing the transition level

D)   in distress

A) Wrong

It is not necessary to squawk an internationally prescribed mode/code when *following a SID (Standard Instrument Departure Route).* Instead, an individual mode/code will be assigned by ATC.

 B) Wrong

It is not necessary to squawk an internationally prescribed mode/code when *flying within controlled airspace.* Instead, an individual mode/code will be assigned by ATC.

 C) Wrong

It is not necessary to squawk an internationally prescribed mode/code when *passing the transition level.* Instead, an individual mode/code will be assigned by ATC.

 D) Correct

An *aircraft in distress* should squawk *A/C 7700* as soon as possible to inform ATC. Additionally, the pilot should inform ATC via radiotelephony.

Other internationally prescribed codes are *A/C7500* for *Hijacking* and *A/C7600* for *radio communication failure.*

 Transpondermode/-code A/C/S7700 zeigt eine Notlage an!

Prüfungsvorbereitung für die Privatpilotenlizenz
Band 8B: Allgemein gültiges Sprechfunkzeugnis (AZF)
2. Auflage 2009

168. A pilot squawking A7500 indicates to the ground station that

A)   he has radio communication failure

B)   his flight is being hijacked

C)   he has a sick person on board

D)   his aircraft is an emergency situation

A) Wrong

The international transponder mode/code for *radio communication failure* is *A/C/S 7600*.

 B) Correct

When a flight is being *hijacked*, the pilot should squawk *A/C/S 7500*.

 C) Wrong

Having a *sick person* onboard, the pilot may declare an *urgency situation* ("PAN PAN").

 D) Wrong

The international transponder mode/code for an *emergency situation* is *A/C/S 7700*.

 Transpondermode/-code A/C/S 7500 zeigt eine Entführung an!

169. Urgency is defined as...

A)   a condition concerning the safety of a person on board or within sight and requiring immediate assistance

B)   a condition concerning the safety of an aircraft or of a person on board, but which does not require immediate assistance

C)   a condition concerning the attitude of an aircraft when intercepting the localizer during an ILS approach

D)   a condition of being threatened by serious and/or imminent danger and of requiring immediate assistance

A) Wrong

Urgency is *also defined* as a condition concerning the *safety of an aircraft!*

 B) Correct

Urgency is defined as a condition concerning the *safety of an aircraft* or of a *person on board*, but which *does not require immediate assistance*. Examples are engine trouble with still running engines, medical problems of passengers or low fuel.

 C) Wrong

Urgency has *no relation to the attitude* of an aircraft.

 D) Wrong

An *emergency* is defined as a condition of being threatened by serious and/or imminent danger and of requiring immediate assistance.

 Eine Situation zur Übermittlung einer Dringlichkeitsmeldung liegt dann vor, wenn die Sicherheit einer Person an Bord oder eines Luftfahrzeuges selbst beeinträchtigt ist, aber eine sofortige Hilfe nicht erforderlich ist.

170. A signal sent by radiotelephony consisting of the spoken words
PAN PAN, preferably spoken three times, means

    A)   the aircraft has a very urgent message to transmit concerning the
            safety of a ship, aircraft or other vehicle

    B)   imminent danger threatens the aircraft and immediate assistance
            is required

    C)   the aircraft is diverting from the route cleared because of a
            thunderstorm and asks for immediate reclearance

    D)   an aircraft on final approach is starting the missed approach
            procedure

 **A) Correct**

A signal sent by radiotelephony consisting of the spoken words *PAN PAN* (e.g. spoken three times) is an *urgency message*, i.e. the aircraft has a very *urgent message to transmit.* An urgency situation results if the *safety of a ship, aircraft or other vehicle is compromised, but does not require help.*

Example: PAN PAN PAN PAN PAN PAN D-GZOF one of two engines out, we perform an immediate safety landing.

 B) Wrong

If *imminent danger threatens the aircraft and immediate assistance is required*, the pilot has to transmit an emergency message (MAYDAY).

 C) Wrong

If the aircraft is *diverting from the route cleared because of a thunderstorm and asks for immediate reclearance*, an urgency message is not required, only a normal request.

 D) Wrong

When starting the *missed approach procedure*, an urgency message is not required.

 PAN PAN sollte bei einer Dringlichkeitsmeldung drei Mal gesprochen werden.

171. An urgency message shall be preceded by the radiotelephony urgency signal:

    A)   PAN PAN, preferable spoken three times

    B)   URGENCY, spoken three times

    C)   MAYDAY, spoken three times

    D)   ALERFA, spoken three times

A) Correct

PAN PAN (preferable spoken three times) indicates an *urgency message*, e.g. engine trouble.

 B) Wrong

The radiotelephony signal *URGENCY* does not exist and must not be used.

 C) Wrong

The radiotelephony signal *MAYDAY* indicates an absolute *emergency*, not an urgency situation.

 D) Wrong

ALERFA means Alerting phase, but is not a spoken radiotelephony signal.

 Dringlichkeitsmedlungen wird die Sprechgruppe PANPAN vorangestellt!

172. Which frequency shall be used for the first transmission of an urgency call?

A)   The air-ground frequency in use at the time

B)   The international emergency frequency

C)   The regional guard frequency

D)   Any frequency at pilot's discretion

A) Correct

Urgency messages should be transmitted on the *air-ground frequency* in use at that time. Changing frequency e.g. to the emergency frequency, may produce delays in transmission.

 B) Wrong

Urgency messages can, but need not be transmitted on the *international emergency frequency.*

 C) Wrong

Urgency messages need not to be transmitted on the *regional guard frequency.*

 D) Wrong

It is *not recommended* to choose a frequency randomly. Who should read your transmission on e.g. 124.775 MHz ?!

 Dringlichkeitsmeldungen sollen und können auf der gerade aktiven Frequenz verbreitet werden.

173. Hearing an urgency message a pilot shall...

A)   acknowledge the message immediately

B)   monitor the frequency to ensure assistance if required

C)   impose radio silence on the frequency in use

D)   change the frequency, because radio silence will be imposed on the frequency in use

A) Wrong

It is not required and may delay further communication to *answer an urgency message*. Therefore, the pilot shall omit this.

 B) Correct

When hearing an urgency message a pilot shall *monitor the frequency to ensure assistance if required*. Examples for urgency messages are bad weather (e.g. no visibility), engine trouble or passenger diseases.

 C) Wrong

It is not absolutely necessary to *impose radio silence on the frequency in use*. If another aircraft requires help, it is allowed and recommended to provide helpful comments or information to the other pilot.

 D) Wrong

Changing the frequency after receiving an urgency message may prohibit ensuring assistance. Therefore, it is recommended to monitor the frequency carefully.

 Nach dem Empfang einer Dringlichkeitsmeldung sollte man die betreffende Frequenz sorgfältig abhören, um dem Luftfahrzeug gegebenenfalls Hilfe anbieten zu können.

174. What is the transponder code for radio communication failure?

A)   6700

B)   7500

C)   7600

D)   7700

A) Wrong

The transponder code *6700* is *not defined* for Germany.

 B) Wrong

The transponder code *7500* indicates *hijacking* (Table 9, page 357)!

 C) Correct

If an aircraft is unable to *establish communication* due to *radio equipment failure*, the pilot should immediately *squawk 7600* to inform ATC (Air Traffic Control).

 D) Wrong

The transponder code *7700* indicates *emergency* (Table 9, page 357)!

SSR Transponder code	Situation
7500	Hijacking
7600	Radio communication failure
7700	Emergency

Table 9: Transponder codes for emergency situations

175. An aircraft is squawking 7600. This indicates:

A)   It is diverting to the alternate aerodrome

B)   It is requesting immediate level change

C)   It is about to make a forced landing

D)   It is unable to establish communication due to radio equipment failure

A) Wrong

For *diverting to an alternate aerodrome*, no special transponder code is defined.

 B) Wrong

For an *immediate request to change the flight level*, no special transponder code is defined.

 C) Wrong

For *making a forced landing*, no special transponder code is defined.

 D) Correct

If an aircraft is unable to *establish communication* due to *radio equipment failure*, the pilot should immediately *squawk 7600* to inform ATC (Air Traffic Control). Other codes are defined for hijacking and emergencies (Table 9, page 357).

 Für Notfälle existieren drei wichtige Transpondercodes: 7500 (Entführung), 7600 (Funkausfall) und 7700 (Notfall)

176. An aircraft station fails to establish radio contact with an
aeronautical station on the designated frequency. What action is
required by the pilot?

A)   Attempt to establish contact on another frequency of the
       aeronautical station

B)   Continue the flight to the destination airport without any
       communication

C)   Return to the airport of departure

D)   Land at the nearest airport without an ATC unit

A) Correct

If contact cannot be established on one frequency, it may be *disturbed* or *not active*. In such a case, the pilot shall try to contact the station on an *alternative (published) frequency*.

 B) Wrong

Continuing the flight without any communication may be severely dangerous.

 C) Wrong

In the given situation, *returning to the airport of departure* is not required.

 D) Wrong

It is not required to *land at the nearest airport without an ATC* (Air Traffic Control) *unit*. Instead, the pilot shall try to contact the station on an alternative (published) frequency.

 Kann eine Bodenstation nicht auf der vorgegebenen Frequenz erreicht werden, sollte man eine alternative Frequenz ausprobieren!

177. What action is required by the pilot of an aircraft station which fails to establish radio contact with an aeronautical station?

A)   Divert to the alternate airport

B)   Squawk mode A code 7500

C)   Land at the nearest aerodrome appropriate to the route of flight

D)   Try to establish communication with other aircraft or aeronautical stations

A) Wrong

It is not required to *divert to another airport.* Instead, the pilot shall try to contact the station on an alternative (published) frequency.

 B) Wrong

Transponder code 7500 indicates hijacking but not radio communication failure (Table 9, page 357).

 C) Wrong

It is not required to *land at the nearest airport.* Instead, the pilot shall try to contact the station on an alternative (published) frequency.

 D) Correct

If contact cannot be established on one frequency, it may be *disturbed* or *not active.* In such a case, the pilot shall try to contact the station on an *alternative (published) frequency* or try to contact other *aircraft* or *aeronautical stations.*

 Kann keine Funkverbindung hergestellt werden, kann man auch probieren andere Luftfahrzeuge bzw. Bodenstationen zu erreichen.

178. A message preceded by the phrase "transmitting blind due receiver failure" shall be transmitted...

A)   on the regional guard frequency

B)   on the international emergency frequency

C)   on the frequency presently in use

D)   to all available aeronautical stations

A) Wrong

It is not necessary to use the *regional guard frequency* for a blind transmission.

 B) Wrong

It is not necessary to use the *international emergency frequency* for a blind transmission.

 C) Correct

Blind transmissions are made when it is not assured that the own radio works sufficiently. These *messages should be sent twice on the designated or presently used frequency* and begin with the phrase "*transmitting blind*".

 D) Wrong

A blind transmission is (hopefully received) by *all aeronautical and all ground stations* listening on the present frequency.

 Blindsendungen werden auf der momentan aktiven Frequenz übermittelt!

179. If all attempts of an aircraft station to establish radio contact with an aeronautical station fail, it shall transmit messages preceded by the phrase:

A)   "Read you one, read you one"

B)   "How do you read?"

C)   "Transmitting blind"

D)   "PAN PAN, PAN PAN, PAN PAN"

A) Wrong

The phrase *"read you one, read you one"* is not common in aviation radio communication. It is only used if communication quality is very low ("example: "D-EZOF read you one"), but is not repeated.

 B) Wrong

The phrase *"How do you read?"* is used to check radio signal quality, but not for blind transmissions.

 C) Correct

Blind transmissions are made when it is not assured that the own radio works sufficiently. These *messages should be sent twice on the designated frequency* and begin with the phrase *"transmitting blind"*.

Example: D-EZOF *transmitting blind* final approach runway 32,

          D-EZOF transmitting blind final approach runway 32

 D) Wrong

The phrase "PAN PAN, PAN PAN, PAN PAN" indicates a *distress message* but may not be used for blind transmissions.

 Blindsendungen beginnen mit der Sprechgruppe "Transmitting blind" und sollten mindestens einmalig wiederholt werden!

180. Blind transmissions shall be made...

A)   only once on the designated frequency

B)   on the emergency frequency only

C)   twice on the designated frequency

D)   during VFR flights only

A) Wrong

Blind transmissions should be made *twice not once.*

 B) Wrong

Blind transmissions can be *made on every frequency.*

 C) Correct

Blind transmissions are made when it is not assured that the own radio works sufficiently. These *messages should be sent twice on the designated frequency.* **Example:**

D-EZOF transmitting blind final approach runway 32,

D-EZOF transmitting blind final approach runway 32

 D) Wrong

Both during VFR and IFR flights blind transmissions may be required and may be made.

181. When transmitting a message preceded by the phrase "transmitting blind due to receiver failure" during an en-route flight, the aircraft station shall also...

A)   enter immediately base leg when approaching the airfield for landing

B)   advise the time of its next intended transmission

C)   land at the nearest airfield/airport

D)   return to the airport of departure

A) Wrong

It is not necessary and additionally dangerous to *enter the base leg immediately*. Instead, the pilot should make a "blind transmission".

 B) Correct

When transmitting a message preceded by the phrase "*transmitting blind due to receiver failure*" (="*Blindsendung*"), thepilot shall also advise the *time of its next intended transmission.*

 C) Wrong

While having a radio communication failure, it is not necessary to *land at the nearest airport immediately*. For the given situation, clear instructions do apply to proceed.

 D) Wrong

While having a radio communication failure, it is not necessary to *return to the airport of departure*. For the given situation, clear instructions do apply to proceed.

182. Under which of the following circumstances shall an aircraft station squawk an internationally prescribed code?

A)   In case of radio communication failure

B)   When entering bad weather areas

C)   When approaching a prohibited area

D)   When flying over desert areas

A) Correct

When having a *radio communication failure*, an aircraft shall squawk 7600 (Table 9, page 357).

 B) Wrong

Entering *bad weather* areas does not require squawking a special code.

 C) Wrong

Approaching a *prohibited area* does not require squawking a special code.

 D) Wrong

Fyling over *desert areas* does not require squawking a special code.

 Für Funkausfall, Entführung und Notfälle existiert jeweils ein definierter Transponder-Code!

183. Which aircraft shall, during radio communication failure, keep a watch for instructions issued by visual signals?

A) VFR flights above clouds

B) Aircraft forming part of the aerodrome traffic at a controlled aerodrome

C) Aircraft entering the traffic pattern of an uncontrolled airport

D) IFR flights when entering a CTR

A) Wrong

It is not possible to see (ground) *lights when flying above clouds*.

 B) Correct

All aircraft forming *part of the aerodrome traffic at a controlled aerodrome* should watch out for instructions issued by visual signals. Especially when having a *radio communication failure*, visual ground signals may be used.

 C) Wrong

An *uncontrolled airport* does not provide ATC and usually has no radar. This implies the person at the facility ("Flugleitung") does not know if an aircraft with radio communication failure arrives. Therefore, it is *not necessary to watch out for instructions issued by visual signals* when entering the traffic pattern of an uncontrolled airport

 D) Wrong

IFR flights are not required to watch out for visual signals, since these flights may take place under instrument weather conditions (nearly no visibility).

 Beim Fliegen in der Platzrunde muss man nach Lichtzeichen Ausschau halten!

Prüfungsvorbereitung für die Privatpilotenlizenz
Band 8B: Allgemein gültiges Sprechfunkzeugnis (AZF)
2. Auflage 2009

184. An IFR flight in IMC encountering radio communications failure, the pilot shall...

A)   leave controlled airspace and continue the flight within uncontrolled airspace

B)   squawk IDENT and proceed to the alternate aerodrome

C)   continue the flight to destination aerodrome

D)   set the transponder to code 7600 and maintain the last assigned speed and level for a period of 7 minutes

A) Wrong

Leaving controlled airspace when flying under IFR conditions may be *deleterious* and *endanger* both aircraft occupants and other aircraft.

 B) Wrong

Squawking IDENT is not necessary in this situation and should be omitted. Additionally, *proceeding to the alternate airport* is only allowed, if *permission* was requested previously and is granted.

 C) Wrong

Continuing an IFR flight with radio communication failure to the destination aerodrome *may cause severe problems* with other aircraft. It is therefore required to squawk a special code (see answer D).

 D) Correct

Radio communication failure during an IFR flight is a *serious problem*. When performing some defined tasks, one may handle this situation. At first, it is required to *squawk 7600* to inform ATC about the radio communication failure (Table 9, page 357). Additionally, the pilot shall *maintain the last assigned speed and level for a period of 7 minutes*.

 Bei Funkausfall Transponder A/C 7600 rasten!

185. An IFR flight in VMC encountering radio communication failure, the pilot is obliged to maintain his last assigned speed and level for a period of 7 minutes. When does this period commence?

A)   at the beginning of radio communications failure

B)   after noticing radio communications failure

C)   at the last contact with ATC

D)   at the time the last assigned level or minimum cruising altitude is reached or at the time the transponder is set to code 7600, whichever is later

A) Wrong

This time period *cannot begin simultaneously* with the radio communication failure, since the pilot may notice this problem by himself some minutes after the beginning.

 B) Wrong

When noticing radio communication failure, the *pilot should act immediately.* The mentioned time period does not begin at this time point.

 C) Wrong

The *last contact with ATC is irrelevant* for this situation. Therefore, the mentioned time period does not begin at this time point.

 D) Correct

On an IFR flight in VMC encountering radio communication failure, the pilot is obliged to *maintain his last assigned speed and level for a period of 7 minutes.* Hereby, the time period starts at

- the time the *last assigned level* or *minimum cruising altitude* is reached or
- at the time the *transponder is set to code 7600*, whichever is later.

186. An IFR flight in IMC encountering radio communication failure, the pilot is obliges to maintain the last assigned speed and level for a period of 7 minutes. What is the pilot supposed to do thereafter?

A)   proceed to an area from where the flight can be continued according to the visual flight rules

B)   divert to the most suitable aerodrome according to the route of flight

C)   adjust level and speed according to the filed flight plan

D)   execute a VMC approach at the nearest suitable aerodrome

A) Wrong

It is not for sure that the pilot identifies such an *area*. There is a *righ risk for collisions* when diverting from the flight plan.

 B) Wrong

It is not assured that the pilot can land (VMC?) at the next airport. Additionally, there is always a *risk of mid-air collisions* when diverting from the filed flight plan in the given situation.

 C) Correct

When having a *radio communication failure at IFR in IMC* (Instrument Meteorological Conditions), the pilot shall *maintain the last assigned speed and level for a perid of 7 minutes*. After that, he shall adjust level and speed according to the filed flight plan. Diversions from the flight plan route have the risk of a collision with other aircraft or terrain.

 D) Wrong

It is not assured that *VMC conditions* exist at the nearest suitable airport!

187. An IFR flight in IMC encountering radio communication failure, the pilot is obliged to maintain the last assigned speed and level for a period of 7 minutes. What is the pilot supposed to do if the minimum IFR cruising altitude is higher than the last assigned level?

A)   In any case maintain last assigned flight level

B)   The pilot shall climb to the minimum IRF cruising altitude

C)   Hold over present position for 7 minutes then continue in accordance with the filed flight plan

D)   Continue immediately in accordance with the filed flight plan

A) Wrong

It is *dangerous to maintain an altitude below the minimum cruising level* in the situation reported. Therefore, the pilot has to climb!

 B) Correct

There is always a risk for *terrain collision* when flying IFR *below the minimum IFR cruising level*. Therefore, a pilot shall *climb to the minimum IFR cruising altitude* when encountering an in-flight radio communication failure.

 C) Wrong

It is *not necessary to hold over the present position*. Instead, the pilot shall climb and continue in accordance with the filed flight plan!

 D) Wrong

The *pilot shall climb* and continue in accordance with the filed flight plan!

188. An IFR flight in IMC encountering radio communications failure, the pilot shall commence descent over the designated navigational aid serving the destination airport (no EAT received):

A)   After holding for 5 minutes in the holding pattern

B)   Commence descent over the IAF at or as close as possible to the estimated time of arrival according to the current flight plan

C)   Without any delay

D)   After 3 minutes, if an expected approach time was not acknowledged

A) Wrong

Holding for 5 minutes in the holding pattern is not necessary. Instead, the pilot shall descent over the initiall approach fix (IAF) possibly at the estimated time of arrival (ETA) as filed in the flight plan.

 B) Correct

When having a *communication failure* during an IFR flight, the pilot shall proceed *as close as possible to the flight plan filed*. This includes commencing *descent over the initial approach fix* (IAF) exactly (if possible) at the *estimated time of arrival* (ETA) as filed in the flight plan.

 C) Wrong

When having a *communication failure* during an IFR flight, the pilot shall proceed *as close as possible to the flight plan filed*. This includes commencing *descent over the initial approach fix* (IAF) exactly (if possible) at the *estimated time of arrival* (ETA) as filed in the flight plan.

 D) Wrong

When having a *communication failure* during an IFR flight, the pilot shall proceed *as close as possible to the flight plan filed*. This includes commencing *descent over the initial approach fix* (IAF) exactly (if possible) at the *estimated time of arrival* (ETA) as filed in the flight plan.

189. An IFR flight in IMC encountering radio communication failure, the pilot shall land, if possible, within...

A)   30 minutes after noticing the failure

B)   20 minutes after leaving the last assigned and acknowledged level

C)   30 minutes after the estimated time of arrival or the last confirmed approach time, whichever is later

D)   15 minutes after vacating the transition layer

A) Wrong

The pilot has to land *within 30 minutes of the ETA* or the *last confirmed approach time*, but not 30 minutes after the failure.

 B) Wrong

The pilot has to land *within 30 minutes of the ETA* or the *last confirmed approach time*, but not 20 minutes after leaving the flight level.

 C) Correct

Encountering a *communication failure* during an IFR flight is a dangerous situation. At first, the pilot shall *squawk 7600* and *proceed on the route as filed in the flight plan*. The pilot has to land *within 30 minutes* after the estimated time of arrival (ETA) or the last confirmed approach time (whichever is laater).

 D) Wrong

The pilot has to land *within 30 minutes of the ETA* or the *last confirmed approach time*, but not 15 minutes after vacating the transition layer (TL).

 Bei einem Funkausfall während eines IFR-Fluges, sollte man spätestens 30 Minuten nach der ETA oder der bestätigten Lande- zeit landen (späterer Zeitpunkt).

190. An IFR flight in VMC encountering radio communication failure the pilot shall...

A)   climb or descend to the cruising level indicated in the flight plan

B)   land at the nearest suitable aerodrome

C)   conduct his flight in accordance with the rules for encountering radio failures in IMC

D)   maintain the altitude last assigned by ATC for a period of 7 minutes, before proceeding to the nearest suitable aerodrome for landing

A) Wrong

Climbing or descending *without previous clearance* may result in a collision and is therefore forbidden.

 B) Wrong

When under IMC, it is not always possible to land at the nearest airport. Instead, the pilot should *squawk 7600 and proceed on the flight plan route*.

 C) Correct

When having a *radio communication failure*, a pilot shall conduct his flight in *accordance with the rules for encountering radio communication failures in IMC* (Instrument Meteorologicaal conditions). This includes squawking 7600 and *maintaining the current flight plan route*.

 D) Wrong

When maintaining the last heading for *7 minutes* without changing course, one may fly into terrain or have a collision with another airplane. 7 minutes are approx. 14 NM when cruising at 120 kt! In addition to this, landing at the nearest suitable airport may result in a deviation from the flight plan route filed and therefore result in a collision.

191. An IFR flight in IMC encountering radio communication failure while under radar vectors, the pilot shall...

A)   squawk 7600 and maintain the heading last assigned by ATC for a period of 3 minutes and then return to the flight path in accordance with the current flight plan

B)   squawk 7600 and proceed in the most direct manner possible to rejoin the current flight plan route no later then the next significant point, taking into consideration the applicable minimum flight altitude

C)   squawk 7600, maintain present heading for 1 minute and thereafter return to the route indicated in the current flight plan on the shortest way

D)   squawk 7600 and maintain present heading for the next 7 minutes and then return to the flight path in accordance with the current flight plan

A) Wrong

Although *squawking 7600* is correct, the procedure given is dangerous. When maintaining the last heading for *3 minutes* without changing course, one may fly into terrain or have a collision with another airplane.

 B) Correct

When encountering a *radio communication failure*, the pilot must *squawk 7600* immediately (Table 9, page 357). Additionally the pilot must:

- *proceed* in the most direct manner possible to
- *rejoin the current flight plan route* no later then the next significant point,
- taking into consideration the applicable *minimum flight altitude*.

 C) Wrong

Although *squawking 7600* is correct, the procedure given is dangerous. When maintaining the last heading for *1 minute* without changing course, one may fly into terrain or have a collision with another airplane.

 D) Wrong

Although *squawking 7600* is correct, the procedure given is dangerous. When maintaining the last heading for *7 minutes* without changing course, one may fly into terrain or have a collision with another airplane. 7 minutes are approx. 14 NM when cruising at 120 kt!

192. In case of a SSR transponder failure occuring after departure of an IFR flight, the pilot shall...

A)   land at the nearest suitable aerodrome for repair

B)   inform the competent ATC unit immediately

C)   squawk 7600

D)   continue the flight in VMC

A) Wrong

It is *not required to land immediately* due to an immediated transponder failure.

 B) Correct

In case of a SSR (Secondary Surveillance Radar) *transponder failure* occurring *after departure*, position tracking is not possible for ATC but is required for *aircraft separation* during IFR flight. In this case, the pilot shall *inform the competent ATC* (Air Traffic Control) unit immediately to gain a *new clearance* which may include continuing the flight without functioning transponder.

 C) Wrong

Squawking 7600 indicates *radio communication failure* instead of transponder failure.

 D) Wrong

It is not possible to *continue the flight in VMC* (Visual Meteorological Conditions).

 Bei einem Ausfall des Transponders muss die zuständige Luftverkehrskontrollstelle (ATC) umgehend informiert werden!

193. In case the transponder fails before the departure for an IFR flight, the pilot shall...

A)   inform FIS for relay to AIS

B)   insert under item 18 of the flight plan "transponder unserviceable"

C)   obtain prior permission by ATC to conduct the flight

D)   inform ATC after departure

A) Wrong

The pilot has to *contact ATC* (e.g. TOWER), *but not FIS* (Flight Information Service).

 B) Wrong

It is *not appropriate to file a special flight plan*. Instead, the pilot has to obtain a special clearance.

 C) Correct

Flying *IFR without* a *functioning transponder is not possible*. Therefore, if a transponder fails before departure for an IFR flight, the *pilot has to obtain a permission* by ATC to conduct the flight.

Only when flying *VFR below 5,000 ft*, a transponder is recommended but not required.

 D) Wrong

It is *not sufficient to inform ATC after departure*, but required to receive a special permission prior to the flight.

 Fällt der Transponder bereits vor dem Abflug aus, muss eine Sonderfreigabe eingeholt werden!

194. What is a pilot expected to do if he reaches the clearance limit with a functioning VHF radio?

A)   Enter the holding pattern and request further clearance

B)   Proceed to the initial approach fix on the shortest way

C)   Proceed to the initial approach fix according to his current flight plan

D)   Continue VFR to destination or alternate aerodrome

A) Correct

When reaching a *clearance limit* (e.g. *IAF*, Initial Approach Fix or a *reporting point*), a pilot must enter the specified *holding pattern* and *hold until a further clearance is obtained*.

 B) Wrong

It is *not allowed to proceed* to the initial approach fix on the shortest way. The pilot has to follow instructions issued by ATC

 C) Wrong

It is *not allowed to proceed* to the initial approach fix on the shortest way. The IAF can be the clearance limit but needs not to be the limit.

 D) Wrong

It is *not allowed to change flight rules* (e.g. IFR → VFR) without any clearance.

195. ATC issues an EAT in any case if the anticipated delay is more than...

    A)   30 minutes

    B)   20 minutes

    C)   15 minutes

    D)   10 minutes

A) Wrong

It is *not 30 minutes* but 20 minutes delay.

 B) Correct

When *flying IFR enroute*, delays may occur in receiving *clearances for the further routing*. This is usually caused by severe traffic. In this case, ATC (Air Traffic Control) *issues an expected approach time* (EAT) if the anticipated *delay is more than 20 minutes*.

 C) Wrong

It is *not 15 minutes* but 20 minutes delay.

 D) Wrong

It is *not 10 minutes* but 20 minutes delay.

196. A pilot receives the clearance to hold over an enroute reporting point until a specified time. This time is called:

A)   Holding time

B)   Expected approach time

C)   Estimated overhead time

D)   Estimated time of arrival

A) Correct

When flying *IFR enroute*, a pilot may be requested to hold over an *enroute reporting point* until a *specified time*. This time-point is called "*holding time*".

The *holding point* may be a navigational aid (e.g. *VOR* or *NDB*) but also a specified *geographical location* in relation to which the position of the aircraft can be reported.

 B) Wrong

The "*expected approach time*" (EAT) is the time issued by ATC at which it is estimated to begin with the approach, i.e. the time at which the initial approach fix (IAF) being *overflown* or the aircraft will *leave the holding point* to complete its approach.

 C) Wrong

The "*estimated overhead time*" is not defined for aviation.

 D) Wrong

The "*estimate time of arrival*" (ETA) is the time at which it is estimated *to land* at the destination airport.

197. If a pilot has to hold over an initial approach fix ATC will issue an...

A)   exact arrival time

B)   estimated elapsed time

C)   estimated time of arrival

D)   expected approach time, if the delay is more than 20 minutes

A) Wrong

The "*exact arrival time*" is not defined for aviation.

 B) Wrong

The "*estimated elapsed time"* (EET) is the time required to proceed from one point to another (e.g. between two VORs).

 C) Wrong

The "*estimated time of arrival"* (ETA) is the time at which it is estimated that the aircraft will arrive over a designated point.

 D) Correct

The "*expected approach time*" (EAT) is the time issued by the ATC at which it is estimated to *begin with the approach*, i.e. the time at which the initial approach fix (IAF) being *overflown* or the aircraft will *leave the holding point/holding pattern* to complete its approach.

198. Which is the frequency-band containing frequencies of the Aeronautical Mobile Service?

A)   108.000 - 117.975 MHz

B)   1,810 - 2850 kHz

C)   11,650 – 13,200 kHz

D)   117.975 - 137.000 MHz

A) Wrong

A *VOR* or *ILS* can be received in the frequency-band between *108.000 and 117.975 MHz.*

 B) Wrong

The *medium frequency band* (MF) with *medium waves* (MW) ranges from *300 to 3,000 KHz* and is used e.g. by NDB or some public radio stations.

 C) Wrong

The *high frequency band* (HF) with *short waves* (SW) ranges from *3 to 30 MHz* (i.e. 3,000 to 30,000 kHz) and is used by public radio stations.

 D) Correct

The *Auronautical Mobile Service* (AMS) used *VHF* (UKW) *radio communication* to facilitate contact between *ground and aircraft* or *aircraft and aircraft* stations. The frequency band used ranges from *117.975 to 137.000 MHz.*

Caution: Only frequencies ranging from 118.000 to 136.975 MHz can be used for communication!

Examples: Mannheim TOWER   118.400 MHz

Speyer INFO   118.075 MHz

   Der Funksprechverkehr im Flugfunk findet auf den Frequenzen 117.975 bis 137.000 MHz statt.

199. To which frequency band belong the frequencies 118.000 - 136.975 MHz of the Aeronautical Mobile Service?

    A)   Very high frequency

    B)   Very low frequency

    C)   Low frequency

    D)   Medium frequency

A) Correct

The *Auronautical Mobile Service* (AMS) uses *VHF* (UKW) *radio communication* to facilitate contact between *ground and aircraft* or *aircraft and aircraft* stations.

The frequency band ranges from *118.000 to 136.975 MHz* and belongs to the VHF (*Very High Frequency*) band. It is characterized by "ultra short waves" ("Ultrakurzwellen", UKW [german]).

Examples: Langen RADAR   129.675 MHz

                  Mannheim ATIS   136.550 MHz

 B) Wrong

The "*very low frequency*" band (VLF) with ultra long waves (SLW) ranges from 3 to 30 kHz and is used for *submarine communication*.

 C) Wrong

The "*low frequency*" band (LF) with long waves (LW) ranges from 30 to 300 kHz and is used by *public radio stations*.

 D) Wrong

The "*medium frequency*" band (MF) with medium waves (MW) ranges from 300 to 3,000 kHz and is used by *public radio stations*.

200. How much is the channel spacing between consecutive frequencies in the VHF band of the Aeronatical Mobile Service?

A)   8.33 kHz / 25 kHz

B)   50 MHz

C)   8.33 kHz / 100 kHz

D)   100 kHz

A) Correct

25 kHz is the correct answer for channel spacing in the VHF radio band for aviation communication below FL 245. That means two different frequencies are spaced by 25 KHz.

In the upper airspace (> FL 245) and in future for IFR flights, channel spacing is reduced because of the lack of frequencies for communication. Frequencies are then only separated by 8.33 kHz.

Examples:	**118,000 MHz**	**(available with 25 kHz spacing)**
	118,010 MHz	(available with 8.33 kHz spacing)
	118,015 MHz	(available with 8.33 kHz spacing)
	118,025 MHz	**(available with 25 kHz spacing)**
	118,035 MHz	(available with 8.33 kHz spacing)
	118,040 MHz	(available with 8.33 kHz spacing)
	118,050 MHz	**(available with 25 kHz spacing)**

 B) Wrong

The given channel spacing of 50 MHz is too broad for the VHF band.

 C) Wrong

Although 8.33 kHz channel spacing will be introduced soon, the given channel spacing of 100 kHz is too broad for the VHF band.

 D) Wrong

The given channel spacing of 100 kHz is too broad for the VHF band.

201. What are the propagation characteristics of VHF?

A)   The waves are reflected at the ionosphere at the height of about
      100 km and reach the earth surface in the form of sky-waves

B)   The waves travel along the surface of the earth and penetrate into
      valleys in a way that topographical obstacles have no influence

C)   Similar to short waves with practically no atmospheric distrubance

D)   Practically straight-line similar to light waves

A) Wrong

Low (LF) and medium (MF) frequency waves are *reflected at the ionosphere at the height of about 100 km* and reach the earth surface in the form of sky-waves, but not VHF waves.

 B) Wrong

Low (LF) and medium (MF) frequency but not VHF waves *travel along the surface of the earth and penetrate into valleys* in a way that topographical obstacles have no influence.

 C) Wrong

VHF waves do not have *the same characteristics* as short waves.

 D) Correct

In contrast to low (LF) and medium (MF) frequency communication (e.g. NDB/Non Directional Beacon), VHF waves *propagate practically straight-line similar to light waves* and *are not reflected* by any obstacles.

 VHF-Wellen breiten sich nur „quasi-optisch" aus und werden – im Gegeensatz zu Kurz-, Mittel- und Langwellen - so gut wie nicht reflektiert!

202. Which phenomena can influence the reception quality of VHF?

A)   The ionosphere

B)   Electrical discharges as they happen frequently in thunderstorms

C)   Day- and night effect

D)   Level of aircraft and terrain elevations

A) Wrong

In contrast to low (LF) and medium (MF) frequency communication (e.g. NDB/Non Directional Beacon), the *ionosphere* does not influence VHF communication significantly.

 B) Wrong

In contrast to low (LF) and medium (MF) frequency communication (e.g. NDB/Non Directional Beacon), *electrical discharges (thunderstorms)* do not influence VHF communication significantly.

 C) Wrong

In contrast to low (LF) and medium (MF) frequency communication (e.g. NDB/Non Directional Beacon), the *day- and night effect* does not influence VHF communication significantly.

 D) Correct

Level of aircraft and *terrain elevations* may influence signal quality in VHF communication significantly (Table 10), since waves propagate practically straight-line similar to light waves.

Factors affecting Very High Frequency communication	Factors affecting low (LF) and medium (MF) frequency communication (e.g. NDB)
• Aircraft at low level • Aircraft far away from the ground station • Aircraft in the radio shadow zone (e.g. of a mountain)	• Day- and night effect • Ionosphere • Electrical discharges/ Thunderstorms • Coastline phenomenon

Table 10: Factors influencing signal quality in VHF, LF, and MF communication.

203. Under which of the following circumstances may you expect a solid reception of the TOWER frequency 118.2 MHz?

A)   Aircraft at low level but far away from the ground station

B)   Aircraft at low level, far away from the ground station, in the radio shadow zone of a hill

C)   Aircraft at high level in the vicinity of the ground station

D)   Aircraft at low level, in the vicinity of the ground station, in the radio shadow zone

A) Wrong

Low level flying and being *far away of a station* decreases reception of a VHF signal dramatically (Table 11).

 B) Wrong

An *aircraft at low level, far away from the ground station* and beeing in the *radio shadow zone* of a hill has a very weak VHF radio reception (signal quality).

 C) Correct

There are three factors for a *solid (good) VHF radio reception* (Table 11). Two of these are given in this answer.

 D) Wrong

Although being in the *vicinity of a ground station* enhances radio reception, both other characteristics decrease reception dramatically.

Solid reception	Weak reception
Aircraft at high level	Aircraft at low level
Aircraft in the vicinity of the ground station	Aircraft far away from the ground station
Optical path between aircraft and station	Aircraft in the radio shadow zone (e.g. of a mountain)

Table 11: Solid versus weak reception in VHF radio communication.

204. The ELBA / ELT transmits on the following frequencies an emergency signal

A)   121.5, 243.0 or 406.0 MHz

B)   121.5 MHz only

C)   243.0 MHz only

D)   119.2 MHz

A) Correct

The *ELT* (Emergency Locator Transmitter, Figure 7) or *ELBA* (Emergency Location Beacon Aircraft) is activated *manually* (if required) or *automatically* (e.g. after a crash) and transmits an *emergency signal* on the following frequencies:

- *121.500 MHz* (civil aviation, civil frequency)
- *243.000 or* (military aviation, military frequencies)
- *406.000 MHz* (future civil frequency)

 B) Wrong

Besides the given civil frequency, the ELBA / ELT additionally transmit *additionally on military frequencies.*

 C) Wrong

Besides the given military frequency, the ELBA / ELT additionally transmit *additionally on a civil frequency.*

 D) Wrong

This is a *completely wrong* frequency.

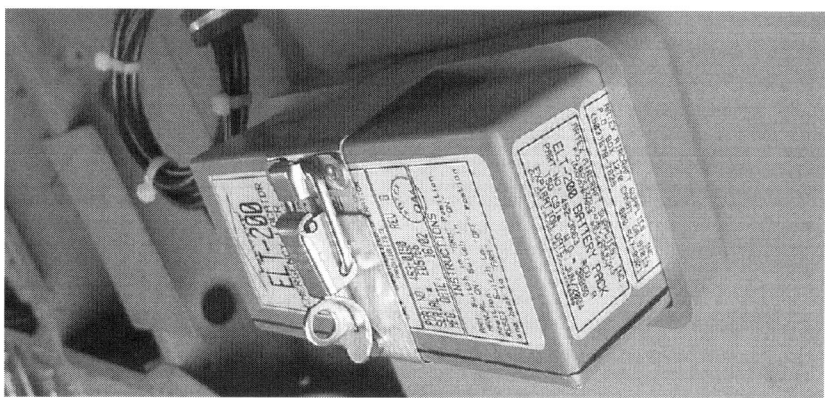

Figure 7: Emergency Locator Transmitter (ELT).

205. Which is the maximum distance at which you might expect solid VHF contact over flat terrain at flight level 100?

A)   About 300 NM

B)   About 30 NM

C)   About 12 NM

D)   About 120 NM

A) Wrong

Actually, the distance is shorter than *300 NM*.

 B) Wrong

Actually, the distance is longer than *30 NM*.

 C) Wrong

Actually, the distance is longer than *12 NM*.

 D) Correct

The *maximum distance* (D_{max}) at which one might expect solid VHF (Very High Frequency) contact depends on the *altitude flown*. To estimate D_{max} the following calculation can be performed:

$$D_{max} \text{ [NM]} = 1.23 \cdot \sqrt{\text{altitude [ft]}}$$

$$= 1.23 \cdot \sqrt{10.000 \text{ ft}}$$

$$= 1.23 \cdot 100$$

$$= 123 \text{ NM} \approx 120 \text{ NM}$$

Caution: This distance is only an estimate and may deviate in reality.

 Mit der o.g. Formel lässt sich die VHF-Reichweite in etwa abschätzen!

8.2.2 AZF Teil II

206. An ATC unit is providing...

A)   aeronautical telecommunication service

B)   air traffic communication service

C)   air traffic information service

D)   air traffic control service

A) Wrong

ATC does not provide *aeronautical telecommunication service*.

 B) Wrong

Air traffic communication service is not provided by an ATC unit.

 C) Wrong

An ATC unit does not provide *air traffic information service*.

 D) Correct

The *ATC unit* (Air Traffic Control) provides *air traffic control service* to aircraft. It is a service provided by ground controllers to *direct aircraft* on the ground and in the air. The primary task is to *separate aircraft* (lateral, vertical and longitudinal separation). Secondary tasks include *ensuring safe*, *orderly* and *expeditious flow of traffic* and *providing information* to pilots, such as weather, navigation information and NOTAMs (Notices to Airmen).

When controllers are responsible for separating some or all aircraft, such airspace is called "*controlled airspace*" in contrast to "*uncontrolled airspace*" where aircraft may fly without the use of the air traffic control system.

207. Air traffic control service will be provided to...

A)   all VFR flights above 5,000 ft MSL

B)   VFR flights in the identification zone

C)   VFR flights at night within airspace G

D)   IFR flights and aerodrome traffic at controlled aerodromes

A) Wrong

ATC service is not provided for VFR flights either *below or above 5,000 ft MSL.*

 B) Wrong

ATC service is not provided for VFR flights either below or above 5,000 ft MSL. Thus, this service is *not available within the identification zone.*

 C) Wrong

ATC service is not provided for VFR flights either below or above 5,000 ft MSL. At night, ATC service is only provided for *NVFR flights within controlled airspace.* Since *airspace G is uncontrolled,* ATC service is not available.

 D) Correct

Air traffic control (ATC) service directs and coordinates airtraffic within a defined area, e.g. *airspace D* (Control Zone, CZ) at controlled airports. Additionally ATC service is provided for all *IFR flights* or for *NVFR flights within controlled airspace.*

 ATC koordiniert IFR-Flüge und alle Flüge an kontrollierten Plätzen.

208. Name the three basic ATC units:

A)   AIS, FIC, TWR

B)   TWR, FIC, APP

C)   APP, ACC, SAR

D)   TWR, APP, ACC

 **A) Wrong**

From the units listed, *AIS* (Aeronatical Information Servive), and *FIC* (Flight Information Center) are not basic *ATC* (Air Traffic Control) units. The FIC is the unit providing flight information servive (*FIS*) and alarming service.

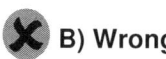 **B) Wrong**

None of the units listed (*TWR*/Tower, *FIC*/ Flight Information Center, and *APP*/Approach) is a basic ATC unit.

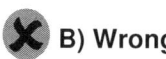 **C) Wrong**

Of the units listed, only *APP*/Approach and *ACC*/Area Control Center are basic ATC units. The *SAR*/Search And Rescue Service is not a basic ATC unit.

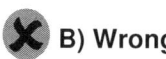 **D) Correct**

TWR (Tower), *APP* (Approiach), and *ACC* (Area Control Center) are the three *basic ATC* (Air Traffic Control) *units*. The *TWR* coordinates traffic within a control zone (CZ), e.g. around an airport. *APP* directs aircraft e.g. from the IAF (Initial Approach Fix) during approach until transferring to the TWR. The *ACC* is a flight information center (FIC) of a flight information region (FIR) and coordinates air traffic within a large area.

209. Information concerning failure of short term duration and/or operational restrictions of ground facilities is published in form of...

A)   AIP supplement

B)   NOTAM

C)   AIC

D)   NfL I

A) Wrong

Only *long term restrictions or changes* are published in the AIP supplement.

 B) Correct

NOTAMs (Figure 8, page 429) are of *very high interest to pilots*, since *important information on airspace restrictions* and *flight routes* are published. A NOTAM is distributed by a *teletype message* (i.e. fax or electronically, e.g. internet or email) and contain e.g. information concerning failure of *short term duration and/or operational restrictions of ground facilities.*

 C) Wrong

The *AIC* (Aeronautical Information Circular) does not publish short term changes.

 D) Wrong

The *NfL* (Nachrichten für Luftfahrer) consists of *part I* and *part II*. In part I long term changes in specific *aeronautical regulations* are published.

210. A NOTAM is distributed by...

A)   an official gazette

B)   a teletype message

C)   no means, since it is of no interest to pilots

D)   radio

A) Wrong

NOTAMS are not distributed by *gazette*.

 B) Correct

NOTAM (Figure 8) is distributed by a *teletype message* (i.e. fax or electronically, e.g. internet or email). NOTAMs are of *very high interest to pilots*, since important information on airspace restrictions and flight routes are published.

 C) Wrong

NOTAMs are of *very high interest to pilots*, since important information on airspace restrictions and flight routes are published.

 D) Wrong

NOTAM is distributed by a teletype message (i.e. fax or electronically, e.g. internet or email) but *not by radio*.

EDKB

```
BULLETIN AERODROME-TYPE ; PREPARED ON:  18.06.02 18:00
FLIGHT RULES    : VFR
INCLUDED NOTAM : LAST 90 DAYS UP TO 02/06/19 18:00

------------ BONN-HANGELAR

EDKB
FROM 02/06/22 06:00 UNTIL 02/06/23 16:30                    E0636/02
JUN 22:
0600-1100 AD PPR FOR ALL ACFT. TEL 02241 202010.
1100-1500 AD CLOSED FOR WING ACFT AND GLIDER FLYING, PPR FOR
HELICOPTER VIA PPR-COORDINATOR TEL 02161 663055 OR FAX 02161 664406.
AFTER 1500 AD PPR FOR ALL ACFT.
JUN 23:
0600-1630 AD CLOSED FOR WING ACFT AND GLIDER FLYING, PPR FOR
HELICOPTER VIA PPR-COORDINATOR TEL 02161 663055 OR FAX 02161 664406.
AFTER 1630 AD REOPENED TO ALL ACFT.
```

Figure 8: NOTAM sample

211. The collection of active NOTAM presented to the pilot by AIS is called:

A)   NfL

B)   Circular Information

C)   Flight plan

D)   Pre-flight Information Bulletin

A) Wrong

The *NfL* (Nachrichten für Luftfahrer) consists of *part I* and *part II*. In part I changes in specific *aeronautical regulations* are published. Part II reports important information and *changes on licensing of both aircrew and aircraft.*

 B) Wrong

The *AIC* (Aeronautical Information Circular) does not represent a collection of NOTAM.

 C) Wrong

A *flight plan* is a special form to report intended flights to the AIS.

 D) Correct

Pilots have to obtain *pre-flight information* from *AIS* (Aeronautical Information Service) for all flights for which a *flight plan has to be filed*. The AIS then provides important information on the flight route, e.g. *NOTAM*, SNOWTAM etc. Accordingly, the pre-flight information bulletin presents a *summary of all active NOTAM* (=NOtice To AirMen).

 Im Pre-flight Bulletin sind alle relevanten und aktuellen NOTAM-Berichte zusammengestellt.

212. The Pre-Flight Information Bulletin presents a summary of all...

A)   active NOTAM

B)   national regulations

C)   international regulations

D)   military activities within the next 24 hours

A) Correct

Pilots have to obtain *pre-flight information* from *AIS* (Aeronautical Information Service) for all flights for which a *flight plan has to be filed*. The AIS then provides important information on the flight route, e.g. *NOTAM*, SNOWTAM etc. Accordingly, the pre-flight information bulletin presents a *summary of all active NOTAM* (=NOtice To AirMen).

 B) Wrong

National regulations are published in the *AIP* (=Aeronatical Information Publication).

 C) Wrong

International regulations are partly published in the *AIP* (=Aeronatical Information Publication) and by *specific aeronatical publications* of other *countries*.

 D) Wrong

Military activities within the next 24 hours may be part of a NOTAM but the pre-flight information bulletin is not a summary of all military activities.

213. Changes of instrument procedures are published by means of...

A)   NfL II

B)   alerting messages (ALR)

C)   AIP supplements or AIP amendments

D)   AIC

A) Wrong

NfL II (Nachrichten für Luftfahrer, Teil II) reports important information and changes on licensing of both aircrew and aircraft.

 B) Wrong

Alerting Messages (ALR) are used when an aircraft is overdue ore missing. These messages are not used to report on changes of instrument procedures.

 C) Correct

In Germany, instrument procedures are published in the *AIP* (Aeronautical Information Publication). *Changes* of these procedures are published in *AIP supplements* or *AIP amendments*. These are available every 4 weeks and published by the „Büro der Nachrichten für Luftfahrer" of the "Deutschen Flugsicherung (DFS)".

 D) Wrong

The *AIC* (Aeronautical Information Circular) does not publish changes of instrument procedures.

 Änderungen der Instrumentenanflugverfahren werden in den AIP-Supplements oder AIP-Ammendments zusammengestellt.

214. Pilots have to obtain pre-flight information from AIS for...

A)   flights for which a flight plan has to be filed

B)   IFR flights only

C)   all flights

D)   all flights operating during night

 **A) Correct**

Pilots have to obtain *pre-flight information* from *AIS* (Aeronautical Information Service) for all flights for which a *flight plan has to be filed*. The AIS then provides important information on the flight route, e.g. *NOTAM*, SNOWTAM etc.

 **B) Wrong**

Pre-flight information has to be obtained by AIS not only for *IFR* (=Instrument Flight Rules) flights but for all flights for which a flight plan is required.

 **C) Wrong**

For *VFR day flights* pre-flight information is usually not required.

 **D) Wrong**

For flights in the traffic circuit (no flight plan required), pre-flight information does not have *to be obtained.*

215. Which are the units of measurement used in flight operation for wind direction and velocity (except for take-off and landing)?

A)   Degrees magnetic and kilometers

B)   Degrees true and kilometers

C)   Degrees magnetic and miles

D)   Degrees true and knots

A) Wrong

Although "*degrees magnetic*" is used for wind direction during take-off and landing, in flight operations except for take-off and landing, "*degrees true*" are used. "*Kilometers*" are not used for wind speed during flight.

 B) Wrong

"*Kilometers*" is not used for wind speed during flight.

 C) Wrong

Although "*degrees magnetic*" is used for wind direction during take-off and landing, in flight operations except for take-off and landing, "*degrees true*" are used. The unit of measurement for in-flight wind speed is "*knots*".

 D) Correct

"Knots" is used as a unit for windspeed for both cruise flight and take-off/landing. For wind direction, "*degrees true*" is used during cruise flight (except for take-off and landing).

 Mit Ausnahme von Start und Landung werden für Windangaben die Einheiten "Grad rechtweisend" und "Knoten" verwendet.

216. Which are the units of measurement used in flight operation for wind direction and velocity for take-off and landing?

A)   Degrees magnetic and knots

B)   Degrees true and kilometers

C)   Degrees magnetic and miles

D)   Degrees true and knots

A) Correct

The *units of measurement* used in flight operation for *wind direction* and *velocity* for take-off and landing are defined as "*degrees magnetic*" (°) and "*knots*" (kt).

 B) Wrong

"*Degrees true*" and "*kilometers*" are never used for wind direction and velocity for take-off and landing.

 C) Wrong

Although "*degrees magnetic*" is used, "*miles*" are not used to describe wind direction and velocity for take-off and landing.

 D) Wrong

Although "*knots*" were used, "*degrees true*" is not used for wind velocity during take-off and landing. Instead, "*knots*" and "*degrees true*" are used during criuse flight.

 Für Start und Landung werden für Windangaben die Einheiten "Grad missweisend" und "Knoten" verwendet.

217. What is the meaning of the designator "A65" on an enroute chart?

A)   Altitude 6500ft

B)   Danger area number six five

C)   Low flying area six five

D)   ATS route A65

A) Wrong

Altitude is presented as a *number only* in the enroute chart (Figure 9, page 475).

 B) Wrong

Danger areas are presented as "ED-D...", e.g. *ED-D65*.

 C) Wrong

Low flying areas usually do not have a abbreviation.

 D) Correct

In the *German enroute chart* (Figure 9, page 475) all airways have an abbreviation. E.g. "A65" means *route "AMBER 65"*.

218. ATC must be informed of a change of the TAS by...

A)   5 % or more

B)   10 % or more

C)   15 % or more

D)   20 % or more

A) Correct

ATC (Air Traffic Control) has to be informed on TAS (True Air Speed) to guarantee adequate separation. If the TAS changes *5% or more*, ATC has to be informed immediately.

 B) Wrong

It is not *10%* but 5% change in TAS.

 C) Wrong

It is not *15%* but 5% change in TAS.

 D) Wrong

It is not *20%* but 5% change in TAS.

 Bei Änderungen der TAS von mehr als 5 % muss ATC informiert werden.

219. In order to conduct an IFR flight "AT FL 200" the aircraft must be equipped with...

A)   one transceiver with 25 kHz channel spacing

B)   one transceiver with 25 kHz and one with at least 50 kHz channel
        spacing

C)   one transceiver with 720 channels

D)   two functioning transceivers with 8.33 kHz channel spacing

A) Wrong

One single transceiver is *not sufficient* to conduct an IFR flight.

 B) Wrong

Transceivers with *50 kHz channel spacing* are not sufficient to conduct a flight according to IFR.

 C) Wrong

One single transceiver (with 720 channels) is not sufficient to conduct an IFR flight.

 D) Correct

To conduct a flight at FL 200 according to IFR, at least *two transceivers* with *8.33 kHz channel spacing* are required.

220. An IFR flight without DME interrogator may be operated...

A)   only with special permission

B)   to and from international airports with RADAR control

C)   if the aircraft is equipped with VOR and ADF

D)   below FL 100, but not to or from international airports

 **A) Correct**

A *DME* (Distance Measuring Equipment) is standard when flying IFR to *measure distances* (e.g. from the current position to the next VOR/DME station) or for non-precision approaches. Performing an IFR flight without DME interrogator is possible *but required special permission.*

 **B) Wrong**

A special permission is required independently from *international airports with RADAR control.*

 **C) Wrong**

For IFR flights all aircraft have to be equipped with *VOR* and *ADF*. Performing an IFR flight without DME interrogator is allowed but *requires special permission.*

 **D) Wrong**

Performing an IFR flight without DME interrogator is possible but *requires special permission.* This has nothing to do with the *FL* or the type of *airport.*

 Einen IFR-Flug ohne DME darf nur mit vorheriger Genehmigung angetreten werden.

221. An IFR flight intending to conduct an instrument approach has to be equipped with LLZ-, GP- and Marker beacon receivers...

A)   during ILS approach

B)   in any case

C)   during SRE approach

D)   during NDB/DME approach

A) Correct

LLZ- (Localizer) receiver, *GP- (Glidepath) receiver*, and *Marker beacon* are required for an *ILS (precision) approach*. With these three receivers, the aircraft can be exactly positioned during an approach.

 B) Wrong

The mentioned receivers are not required in any case. For a non-precision *VOR/DME* or *NBD/DME* approach, a glidepath indicator is unnecessary.

 C) Wrong

For a non-precision *SRE approach*, a glidepath indicator is unnecessary.

 D) Wrong

For a non-precision *NBD/DME approach*, a glidepath indicator is un-necessary.

222. By what time at the latest prior to EOBT will AIS/IFPS normally accept a flight plan?

A)   3 hours

B)   2 hours

C)   1 hour

D)   0.5 hours

A) Wrong

The latest time prior to EOBT is only 1 hour *not 3 hours.*

 B) Wrong

The latest time prior to EOBT is only 1 hour *not 2 hours.*

 C) Correct

The latest time for the AIS (Aeronatical Information Service) to accept a flight plan is *1 hour* prior to *EOBT* (Estimated Off-Block Time). A flight plan can be filed by phone, fax or the internet.

 D) Wrong

The latest time prior to EOBT is only 1 hour, not 0.5 hours.

 Ein Flugplan muss mindestens eine Stunde vor Abflug (maximal drei Tage vor Abflug) aufgegeben werden.

223. A pilot intends to fly IFR via ATS route "G1" till "ABC" VOR and thereafter VFR. What is the correct entry in his flight plan under item 15 "route"?

A)   VFR ABC G1

B)   N0120 VFR ABC G1

C)   G1 N0120/ABC VFR

D)   G1 ABC VFR

A) Wrong

The presented answer (*VFR ABC G1*) indicates the opposite: Flying VFR and when overflying ABC routing is G1.

 B) Wrong

The airspeed (*N0120*) is not reported under item 15 "route" when changing from IFR to VFR. Therefore, this answer cannot be correct.

 C) Wrong

The airspeed (*N0120*) is not reported under item 15 "route" when changing from IFR to VFR. Therefore, this answer cannot be correct.

 D) Correct

The presented answer is correct, since actual routing is *G1*. When overflying *ABC*, the flight changes to *VFR*. Therefore, the answer has to be: *G1 ABC VFR*.

 Unter „Item 15" wird im Flugplan die zu fliegende Route einge-tragen!

224. A pilot intends to fly VFR till "XYZ" VOR and thereafter IFR. What is the correct entry in his flight plan under item 15 "route"?

A) VFR XYZ IFR

B) XYZ/N0120F080 IFR

C) 0120/XYZF080 IFR

D) F080 XYZ IFR

A) Wrong

Reporting only *VFR XYZ IFR*, when changing from VFR to IFR is not sufficient.

 **B) Correct**

When flying VFR and changing to IFR, a pilot must file a *"Z"* flight plan. Additionally he has to enter a correct entry in *item 15 "route"*. This consists of a *reference point* (XYZ) followed by a *"/"* as well as *airspeed* and *altitude or flight level* (N0120F080). After this point the flight is commenced *IFR*.

Therefore, the correct answer is: XYZ/N0120F080 IFR.

 C) Wrong

For the airspeed the letter *"N"* is required to indicate Nautical Miles per hour (i.e. kt, knots).

 D) Wrong

The *sequence* is not correct and the *airspeed* is missing.

225. What is the correct entry in an IFR flight plan for a flight level change over Leipzig VOR (LEG)?

A)   ... A101 N0250/F150 LEG G98 ...

B)   ... A101 LEG/N0250F150 G98 ...

C)   ... A101 F150/N0250 LEG G98 ...

D)   ... A101 LEG/F150 G98 ...

A) Wrong

Although the three required items (see answer B) are presented, the *sequence* is incorrect (A101 N0250/F150 LEG G98).

 B) Correct

When changing a flight level, three items have to be reported in the flight plan in the following sequence:

- *Reference point/*
- *Airspeed (kt)*
- *New flight level (FL)*

In addition, the routing has to be reported: A101 ....G98. Combining these data, the correct answer is: ... *A101 LEG/N0250F150 G98 ...*

 C) Wrong

The *sequence* of the data is incorrect (A101 F150/N0250 LEG G98). *Airspeed* has to be reported before the intended flight level.

 D) Wrong

In this answer (A101 LEG/F150 G98), *airspeed* has to be reported also.

226. Which letter is used in the flight plan to indicate that the flight commences in accordance with IFR and subsequently changes to VFR?

A)   Z

B)   Y

C)   I

D)   V

A) Wrong

A "*Z*" in the flight plan indicates that a flight commences in accordance with *VFR* and subsequently *changes to IFR* later on.

 **B) Correct**

A "*Y*" in the flight plan indicated that a flight commences in accordance with *IFR* and subsequently *changes to VFR*.

 C) Wrong

The "I" in the flight plan indicates *IFR traffic*.

 D) Wrong

The "V" in the flight plan indicated *VFR traffic*.

 Ein Y-Flugplan besagt, dass der Flug nach IFR begonnen wird und später VFR fortgesetzt wird.

227. How many minutes after EOBT is the flight plan automatically cancelled by ATC, if start-up or taxi instructions have not been requested?

A)   30 minutes

B)   60 minutes

C)   90 minutes

D)   120 minutes

A) Wrong

The time period is *longer*: 60 minutes.

 B) Correct

Air traffic control (ATC) cancels a flight plan exactly *60 minutes* after *EOBT* (Estimated Off-Block Time) if start-up or taxi *instructions have not been requested*.

 C) Wrong

The time period is *shorter*: 60 minutes.

 D) Wrong

The time period is *shorter*: 60 minutes.

 Ein Flugplan wird automatisch von ATC gelöscht, falls nicht innerhalb von 60 Minuten nach der EOBT „start-up" oder Roll-anweisungen beantragt werden.

228. An IFR training flight shall be marked in item 8 of the flight plan by using the letters:

A)   IX

B)   VG

C)   IN

D)   VN

A) Correct

The letters "*IX*" in the item 8 of a flight plan indicate *IFR (I)* traffic under *training conditions (X)*.

 B) Wrong

The letters "*VG*" indicate General Aviation VFR traffic.

 C) Wrong

The letters "*IN*" are not defined.

 D) Wrong

The letters "*VN*" indicate VFR traffic at night, i.e. NVFR.

 Die Abkürzung "IX" im "Item 8" des Flugplanes bedeutet, dass es sich um einen IFR-Trainingsflug handelt.

229. What is the meaning of the term "flight level"?

A)   A pressure level based on regional QNH

B)   A level in the atmosphere for vertical separation which is determined by setting the altimeter to local QNH

C)   A level in the atmosphere for vertical separation which is determined by setting the altimeter to local QFE

D)   A level in the atmosphere for vertical separation which is determined by setting the altimeter to 1,013.2 hPa

A) Wrong

Although the flight levels (FL) are based on *pressure levels*, it is not the *regional QNH* but the *standard pressure* (1,013.25 hPa) which defines the FL.

 B) Wrong

It is not the *local QNH* but the *standard pressure* (1,013.25 hPa) which defines the FL.

 C) Wrong

It is not the *local QFE* but the *standard pressure* (1,013.25 hPa) which defines the FL.

 **D) Correct**

A *flight level* (FL) is defined as a *level in the atmosphere* for *vertical separation* of aircraft (e.g. between VFR and IFR traffic) determined by setting the *altimeter to 1,013.2 hPa* (standard pressure).

230. Except for take-off and landing the minimum safe height for IFR flights is at least:

A)   500 ft above the highest obstacle located within 8 km of the aircraft´s position

B)   1,000 ft above the highest obstacle located within 5 km of the aircraft´s position

C)   1,500 ft above the highest obstacle located within 8 km of the aircraft´s position

D)   1,000 ft above the highest obstacle located within 8 km of the aircraft´s position

A) Wrong

The minimum safe height is *not 500 ft* but 1,000 ft above the highest obstacle located within 8 km of the aircraft position.

 B) Wrong

The minimum safe height is 1,000 ft above the highest obstacle located within 8 km but *not 5 km* of the aircraft position.

 C) Wrong

The minimum safe height is *not 1,500 ft* but 1,000 ft above the highest obstacle located within 8 km of the aircraft position.

 D) Correct

The *minimum safe height* is defined as the minimum height for IFR flights. It is *1,000 ft* above the highest obstacle located *within 8 km* of the aircraft position

 Die MSA beträgt bei IFR-Flügen 1.000 ft über dem höchsten Hindernis in einem Umkreis von 8 km.

231. The IFR minimum on ATS routes is expressed in:

A)   Flight level only

B)   Altitude and/or flight level

C)   Elevation

D)   Height

A) Wrong

Although *flight level* (FL) is one of the *IFR minima on ATS routes*, "*altitude*" is also one possible minimum.

 **B) Correct**

The *IFR minimum on ATS routes* is expressed in *altitude* (below 5,000 ft) or *flight level* (FL; above 5,000 ft).

 C) Wrong

Only *buildings* (e.g. antenna) or *geographical points* (e.g. hill) can have an elevation above the ground.

 D) Wrong

Only *buildings* (e.g. antenna) or *geographical points* (e.g. hill) can have an elevation above the ground.

232. Within how many miles radius around a specified navigational aid does the minimum sector altitude provide 1,000 ft obstacle clearance?

A)   30 NM

B)   25 NM

C)   20 NM

D)   15 NM

A) Wrong

It is 25 NM, *not 30 NM*!

 B) Correct

It is defined that the *minimum sector altitude* (MSA) provides *1,000 ft obstacle clearance* within a *25 NM radius* around a specified navigational aid. In other words, in a radius of 25 NM the next obstacle is at least 1,000 ft lower.

 C) Wrong

It is 25 NM, *not 20 NM*!

 D) Wrong

It is 25 NM, not *15 NM*!

233. Altitude information of route segments on German enroute charts indicates the...

A)   minimum safe height for IFR flights

B)   minimum reception altitude for radio navigational aids

C)   minimum IFR cruising altitude

D)   lower limit of the controlled airspace

A) Wrong

The *minimum safe height* for IFR flights is not published, since height refers to the ground.

 B) Wrong

The *minimum reception altitude for radio navigational aids* is not published on German enroute charts.

 C) Correct

Altitude information of route segments on *German enroute charts* indicates the *minimum IFR cruising altitude* (Figure 9). The numbers are given along the published route and represent the lowest altitude available for IFR flights.

 D) Wrong

The *lower limit of the controlled airspace* is not published on German enroute charts.

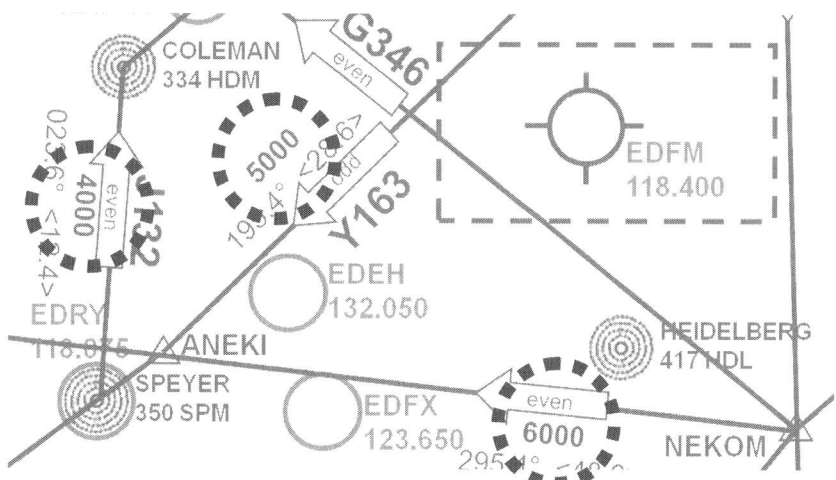

Figure 9: Sample of a German Enroute Chart. Minimum IFR cruising altitudes are marked (not for navigational use since a sample!).

Prüfungsvorbereitung für die Privatpilotenlizenz
Band 8B: Allgemein gültiges Sprechfunkzeugnis (AZF)
2. Auflage 2009

234. What is the minimum vertical separation to other IFR flights above FL 290?

A)   2,000 m

B)   1,000 ft up to FL 410 thereafter 2,000 ft

C)   1,000 ft

D)   500 ft

A) Wrong

Vertical separation is *never 2,000 m.*

 B) Correct

The *minimum vertical separation* to other IFR flights *up to FL 410* is 1,*000 ft* and thereafter 2,000 ft.

 C) Wrong

Vertical separation is *1,000 ft between IFR and IFR flights or VFR and VFR for flights up to FL 410.*

 D) Wrong

Vertical separation is *500 ft between VFR and IFR flights up to FL 410.*

 Der vertikale Abstand zwischen IFR-Flügen beträgt 1.000 ft bis FL 410 und 2,000 ft über FL 410.

Prüfungsvorbereitung für die Privatpilotenlizenz
Band 8B: Allgemein gültiges Sprechfunkzeugnis (AZF)
2. Auflage 2009

235. What is the general minimum vertical separation to other IFR flights below FL 410?

A)   2,000 ft

B)   1,000 ft

C)   1,000 m

D)   500 ft

A) Wrong

The *minimum vertical separation* to other IFR flights *above FL 410* is *2,000 ft.*

 B) Correct

Vertical separation is *1,000 ft between IFR and IFR traffic or VFR and VFR traffic for flights up to FL 410.*

 C) Wrong

Vertical separation is *never 1,000 m.*

 D) Wrong

Vertical separation is *500 ft between VFR and IFR flights up to FL 410.*

 Unter FL 410 beträgt die Separierung zwischen IFR-Flügen 1.000 ft (und 500 ft zu VFR-Flügen).

236. Prior to departure the pilot will be given enroute clearance including the departure route. The cruising altitude is normally not covered by the clearance. To which initial altitude has the pilot to climb after take-off?

A)   Climb so, as to leave airspace E as soon as possible

B)   To the minimum safe height for IFR flights

C)   The initial altitude is at the pilot's discretion

D)   To the initial altitude stated in the SID

A) Wrong

It is not allowed *to climb as soon as possible*, since other traffic might be on a *collision course*.

 B) Wrong

On request, it is allowed to *descent to the minimum safe altitude*. This is the lowest altitude with radar coverage available for IFR flights.

 C) Wrong

It is not allowed to climb to an *altitude at the pilot´s discretion*, since other traffic might be on a *collision course*.

 D) Correct

The *cruising altitude* is not covered by the *enroute clearance*. After take-off, the pilot has to climb to the *initial altitude stated in the SID* (Standard Instrument Departure chart). This altitude may be e.g. 5,000 ft

237. The transition level is determined on the basis of...

A)   transition altitude and QNH

B)   transition height and QNH

C)   transition altitude and QFE

D)   transition height and QFE

A) Correct

The *transition level* is the lowest *flight level (FL)* available for use above the *transition altitude*. The transition layer has a vertical dimension of at least 1,000 ft.

The *transition level* is determined on the basis of *transition altitude and QNH.*

 B) Wrong

The *transition height* is not used to determine the transition level.

 C) Wrong

Although *transition altitude* is used, QNH instead of *QFE* must be used to determine the transition level.

 D) Wrong

Both *QFE* and *transition height* are not used to determine the transition level.

 Das "transition level" bezieht sich auf die "transition altitude" und das QNH!

238. The vertical dimension of the transition layer must be at least:

A)   1,000 ft

B)   500 ft

C)   1,500 ft

D)   2,000 ft

 **A) Correct**

The *transition layer* is the airspace *between* the *transition level* (upper limit) *and* the *transition altitude* (lower limit). In Germany, the transition altitude is *5,000 ft* and the transision *level depends on the current QNH*. Hereby, the vertical dimesion of the *transition layer must be at least 1,000 ft*.

B) Wrong

The vertical dimesion of the transition layer must be at least 1,000 ft.

C) Wrong

The vertical dimesion of the transition layer must be at least 1,000 ft.

D) Wrong

The vertical dimesion of the transition layer must be at least 1,000 ft.

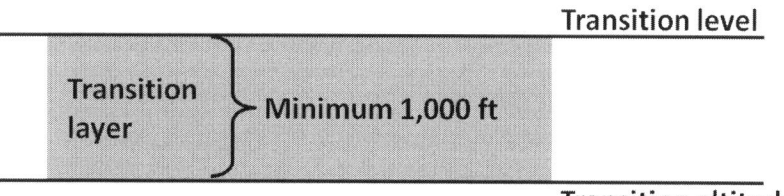

Figure 10: Transition attitude, transition layer, and transition level

239. The published transition altitude is 5,000 ft, the present QNH is 1,005 hPa. What is the transition level?

A)   FL 50

B)   FL 60

C)   FL 70

D)   FL 80

A) Wrong

The transion level *only be below FL 60 if the QNH is above 1,050 hPa.*

 B) Wrong

The transition level is at *FL 60 if the QNH is between 1,014 and 1,050 hPa.* This does not apply to this situation.

 C) Correct

The *transition level* is the lowest *flight level (FL)* available for use above the *transition altitude*. The spacing between these two is called the transition layer. Depending on the transition altitude and the current QNH, the width of the transition layer can change. Below are the values for a transition altitude of 5,000 ft: (Figure 10, page 485).

QNH	TL
> 1,050	50
1,014 – 1,050	60
978 – 1,013	70
< 978	80

 D) Wrong

The answer given is *too high.*

240. The published transition altitude is 5,000 ft, the present QNH is 1,015 hPa. What is the transition level?

A)   FL 50

B)   FL 60

C)   FL 70

D)   FL 80

A) Wrong

The transion level *only be below FL 60 if the QNH is above 1,050 hPa*.

 B) Correct

The *transition level* is the lowest *flight level (FL)* available for use above the *transition altitude*. Here, the width of the transition level is always at least 1,000 ft (Figure 10, page 485).

If the *QNH is higher than standard pressure* (i.e. 1,013.25 hPa), the *transition level is* FL 60.

 C) Wrong

The *transition level moves up if the QNH is lower than standard* pressure (i.e. QNH < 1,013.25 hPa).

 D) Wrong

The *transition level moves up if the QNH is lower than standard* pressure (i.e. QNH < 1,013.25 hPa).

241. The published transition altitude is 5,000 ft, the present QNH is 977 hPa. What is the transition level?

A)   FL 50

B)   FL 80

C)   FL 70

D)   FL 60

A) Wrong

The transition level is only in FL 50, if the actual pressure is above 1,050 hPa).

 B) Correct

The *transition level* is the lowest *flight level (FL)* available for use above the *transition altitude*. The spacing between these two is called the transition layer. Depending on the transition altitude and the current QNH, the width of the transition layer can change. Below are the values for a transition altitude of 5,000 ft: (Figure 10, page 485).

QNH	TL
> 1,050	50
1,014 – 1,050	60
978 – 1,013	70
< 978	80

 C) Wrong

Although it is correct that the transition level is higher than standard, *FL 70 is incorrect.*

 D) Wrong

Although it is correct that the transition level is higher than standard, *FL 60 is incorrect.*

242. An aircraft on an IFR flight at FL 100 approaches an aerodrome for landing. The QNH given is 1,018 hPa, the transition altitude is 5,000 ft. When shall the pilot change the altimeter setting to QNH? When...

A)   leaving FL 100

B)   passing FL 50

C)   passing FL 70

D)   passing FL 60

A) Wrong

The altimeter setting should be changed when passing the transition level. *FL 100 is not the transition level* in this situation.

 B) Wrong

The altimeter setting should be changed when passing the transition level. *FL 50 is the transitions altitude not the transition level* in this situation.

 C) Wrong

The altimeter setting should be changed when passing the transition level. *FL 70 is not the transition level* in this situation.

 D) Correct

The altimeter setting should be changed when passing the transition level (Figure 10, page 485).

If the QNH is higher than standard (i.e. 1,013.25 hPa), the transition level remains at FL 60 and the thickness of the transition layer is more than 1,000 ft. Therefore, FL 60 is the correct answer.

243. The pilot of an arriving IFR flight shall change altimeter setting from 1,013.2 hPa to QNH during the descent...

A)   when passing the transition altitude

B)   when commencing descent

C)   after having passed the transition layer

D)   when passing the transition level

A) Wrong

When arriving at the *transition altitude* on an IFR flight, the altimeter must *already be set* on QNH.

 B) Wrong

The *altimeter setting* should be changed to from standard pressure to QNH when flying IFR and *passing the transition level.*

 C) Wrong

After having passed the transition layer, the aircraft reaches the transition altitude. This is *too low* to set the altimeter to QNH!

 **D) Correct**

When flying IFR, the *altimeter setting* should be changed to QNH *from standard pressure* while *passing the transition level* (Figure 10, page 485 and Figure 11).

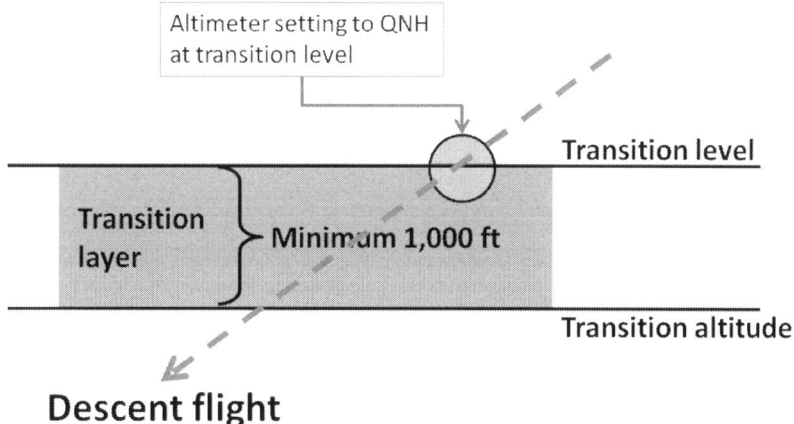

Figure 11: Altimeter setting during descent flight

244. The pilot of a departing IFR flight shall change altimeter setting from QNH to standard altimeter setting 1,013.2 hPa when...

A)   passing transition level

B)   reaching transition level

C)   passing transition altitude

D)   leaving the transition layer

A) Wrong

The altimeter setting should be changed to QNH *during descent* when passing the transition level.

 B) Wrong

The altimeter should never *be changed when reaching the transition level* during climb.

**C) Correct**

During the *climb after departure*, the pilot has to change the *altimeter setting* from *QNH to standard pressure* (1,013.25 hPa) when passing the *transition altitude* (Figure 12).

 D) Wrong

The altimeter should *never be changed when passing or leaving the transition layer.*

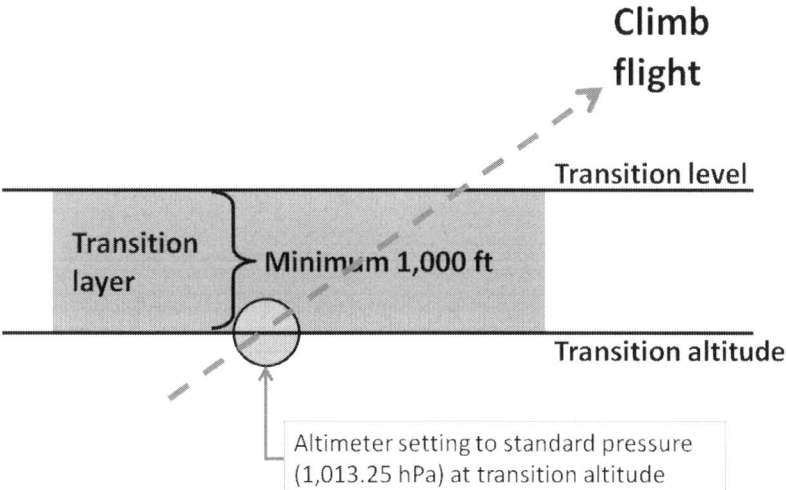

Figure 12: Altimeter setting during climb flight

Prüfungsvorbereitung für die Privatpilotenlizenz
Band 8B: Allgemein gültiges Sprechfunkzeugnis (AZF)
2. Auflage 2009

245. Name the speed limitation applicable within airspace C for an IFR flight:

A)   Maximum 250 kt ISA

B)   No speed limitation

C)   Maximum 250 kt TAS

D)   250 kt IAS below FL 100

A) Wrong

A speed limitation of 250 kt IAS applies to *VFR traffic below FL 100.*

 B) Correct

For IFR traffic *no speed limit* exists when flying in airspace C.

 C) Wrong

There is no maximum speed defined for *an IFR flight.*

 D) Wrong

A speed limitation of 250 kt IAS applies to *VFR traffic below FL 100.*

 IFR-Flüge sind in ihrer Fluggeschwindigkeit in Luftraum C nicht limitiert.

246. Name the lower and upper limits of the german UIR:

A)   FL 245 - unlimited

B)   GND - FL 245

C)   FL 245 - FL 460

D)   GND - unlimited

A) Correct

The German *UIR* (Upper flight Information Region) has a lower limit (i.e. Lower Information Region) but no upper limit. For Europe, different UIR do exist covering more than one country. In Germany, a total of 3 regions do exist.

 B) Wrong

The German Flight Information Region (*FIR*) covers *GND* (GrouND) to *FL 245*.

 C) Wrong

Although the lower limit of UIR is correct, the upper limit ranges up to "*unlimited*".

 D) Wrong

Both Upper and Lower Information region do cover *GND – unlimited*.

247. Flights within airspace E will be separated as follows:

A)   IFR from VFR

B)   VFR from IFR

C)   IFR from IFR

D)   all aircraft

A) Wrong

Flights within *airspace E* are only separated when flying *IFR*. Therefore, IFR traffic cannot be separated from VFR traffic.

 B) Wrong

Flights within *airspace E* are only separated when flying *IFR*. Therefore, VFR traffic cannot be separated from IFR traffic.

 C) Correct

Flights within *airspace E* are only separated *when flying IFR*. This means aircrafts *receive instructions to avoid collision* or information on other traffic.

 D) Wrong

Flights within *airspace E* are only separated when flying *IFR*. Therefore, all traffic cannot be separated.

 In Luftraum E werden nur IFR von IFR-Flügen separiert.

248. Which kind of flights are not permitted within airspace G?

A)   IFR

B)   VFR

C)   Glider flights

D)   Military exercise flights

A) Correct

Airspace G is a completely *uncontrolled airspace* without *radar guidance.* Therefore, flights with required *radar coverage* and *radar guidance* are not possible and not allowed. From the answers given, only *IFR flights* require radar guidance.

 B) Wrong

VFR flights are permitted within *airspace G* (but not in airspace A and B).

 C) Wrong

Glider flights are permitted in uncontrolled *airspace G,* but not in airspace C, B, and A.

 D) Wrong

Military exercise flights are permitted in both controlled and uncontrolled airspace. Therefore, they are allowed in airspace G.

 Im (unkontrollierten) Luftraum G dürfen (in Deutschland) keine IFR-Flüge durchgeführt werden!

249. Flights below FL 100 within airspace D and E are not permitted to exceed an indicated airspeed of...

A)   250 kt except aircraft, which have to be operated at a higher airspeed because of their specific performance characteristics

B)   200 kt

C)   150 kt

D)   300 kt

A) Correct

For flights *below FL 100* within *airspace D and E* a *maximum indicated airspeed* (IAS) is set to prevent *accidents* and *collisions*. Aircraft are not permitted to exceed an indicated airspeed of *250 kt*. This maximum IAS may only be exceeded by aircraft *which have to be operated at a higher airspeed* because of their specific performance characteristics.

 B) Wrong

The maximum IAS is 250 kt – *not 200 kt.*

 C) Wrong

The maximum IAS is 250 kt – *not 150 kt.* Many single-engine aircraft (e.g. PIPER Malibu, SOCATA TB20) may be faster than 150 kt.

 D) Wrong

The maximum IAS is 250 kt – *not 300 kt.*

250. A pilot intends to cancel his IFR flight at the minimum IFR cruising altitude, descend below controlled airspace and continue VFR. Which are the required weather minima?

A)   Flight visibility at least 3 km, vertical distance from clouds 1,000 ft

B)   Flight visibility 8 km, clear of clouds

C)   Flight visibility 8 km, horizontal distance from clouds 1.5 km

D)   Flight visibility at least 3 km, clear of clouds, visual contact to the
      ground

A) Wrong

Although a *visibility of 3 km* is enough, the aircraft must remain *clear of clouds* and the pilot has to assure *visual contact to the ground*.

 B) Wrong

It is not required to have a visibility of *8 km*. This only applies for *airspace C or E*.

 C) Wrong

Both *visibility* and *distance from clouds* are incorrect in the answer presented.

 D) Correct

When *changing flight rules* from IFR to VFR, *special weather conditions* do apply. In the present situation, it is necessary to observe *minimum weather values* for the airspace intended to enter. In this situation it is required to have a visibility of *3 km*. Additionally, the aircraft must remain *clear of clouds* and the pilot has to assure *visual contact to the ground*.

251. A pilot departing on a VFR flight (Z flight plan) and intending to change flight rules to IFR at the minimum IFR cruising altitude, has to observe the following minimum values for flight visibility/ distance from clouds:

A)   1.5 km / clear of clouds

B)   1.5 km within "G" and 3 km within "E" / 1,000 ft vertically

C)   3 km within airspace G and E / clear of clouds

D)   1.5 km within airspace G, 5 km within airspace E / 1.5 km horizontally

A) Correct

When departing on a VFR flight, it is mandatory to have sufficient minimal wether conditions for *airspace G*. These are at least a *visibility of 1.5 km and to remain clear of clouds*.

 B) Wrong

It is not required to have a visibility of 3 km *but 1.5 km* within airspace G when changing flight rules.

 C) Wrong

Although it is mandatory to *stay clear of clouds*, a *visibility of 1.5 km* is sufficient for the situation reported.

 D) Wrong

It is not necessary to have a visibility of 5 km within *airspace E* when changing flight rules.

252. A pilot on an IFR flight (Y flight plan) intends to change flight rules to VFR at the minimum IFR cruising level. Until leaving the controlled airspace he has to observe the following minimum values for flight visibility:

A)   3 km

B)   5 km

C)   8 km

D)   1.5 km

A) Correct

When *changing flight rules* from IFR to VFR, *special weather conditions* are required. In the present situation, it is necessary for the piot to observe *minimum weather values* for the airspace he intends to enter. In this situation it is required to have a visibility of *3 km*. Additionally, the aircraft must remain *clear of clouds* and the pilot has to assure *visual contact to the ground*.

 B) Wrong

Not 5 km but 3 km is the required visibility in this situation.

 C) Wrong

It is not necessary to have a *visibility of 8 km*. This is only required for VFR flights within *airspace E*.

 D) Wrong

A visibility of *1.5 km* is only required for *airspace G*. In the situation reported, the pilot can not be in airspace G, since the *minimum IFR cruising level* is at least in airspace E.

253. During a flight with an intended change of flight rules the VFR part of this flight shall generally be conducted in such a way, that...

A)   in airspace E the pilot has a flight visibility of at least 5 km and the aircraft maintains a horizontal distance from clouds of at least 1.5 km and a vertical distance of at least 1,000 ft

B)   in airspace E the pilot has a flight visibility of at least 8 km and the aircraft maintains a horizontal distance from clouds of at least 1.5 km and a vertical distance of at least 1,000 ft

C)   in airspace E the pilot has a flight visibility of at least 8 km and the aircraft remains clear of clouds

D)   in airspace E the pilot has a flight visibility of at least 5 km, the aircraft remains clear of clouds and visual contact to the ground is granted

A) Wrong

The visibility in airspace E has to be *at least 8 km* but not 5 km!

 B) Correct

When flying under VFR (*Visual Flight Rules*), it is essential to follow *required minima*. The minima for airspace E are as follows:

- *visibility* of at least 8 km
- *horizontal distance* from clouds of at least 1.5 km and
- *vertical distance* of at least 1,000 ft

 C) Wrong

Remaining clear of clouds is mandatory when flying in *airspace G* (*uncontrolled*) but not in airspace E (controlled). Here it is required to maintain a *horizontal distance* from clouds of at least 1.5 km and a *vertical distance* of at least 1,000 ft.

 D) Wrong

The visibility in airspace E has to be at least *8 km* but not 5 km! *Remaining clear of clouds* and having *visual contact to the ground* are mandatory when flying in airspace G (uncontrolled), but not airspace E (controlled).

254. The start-up permission shall only be requested by the pilot, when it is guaranteed that aircraft can start the engines after the permission has been issued...

A)   within 10 minutes

B)   within 5 minutes

C)   immediately

D)   within 20 minutes

A) Wrong

The time period of *10 minutes* given in the answer is *too long*. It is required to start the engines within 5 minutes!

 B) Correct

Start-up permission is granted by *ATC* (Air Traffic Control) and shall be *requested by the pilot* when ready. The pilot has to guarantee, that all *engines are started within 5 minutes* after obtaining permission.

 C) Wrong

It is not required to start the engines after permission *immediately*. In fact, *5 minutes* are sufficient.

 D) Wrong

The time period of *20 minutes* given in the answer is *too long*. It is required to start the engines within 5 minutes!

 Nach einem "Start-up request" sollen die Triebwerke innerhalb von 5 Minuten gestartet sein.

255. For the regulation of taxiing aircraft under all-weather conditions, CAT II/III stop bars have been established...

A)   for the safety of traffic on taxiways and on runways

B)   for the safety of taxiing aircraft

C)   for the safety of traffic on the runways

D)   to ensure a fluent traffic on the apron

A) Correct

If the red lights of a stop bar are switched on, the pilot has to *hold position* under all circumstances. It is not allowed to pass this sign due to *safety reasons*. CAT II/III stop bars have been established for the *safety of traffic on taxiways and on runways*.

 B) Wrong

In addition to the safety of taxiing aircraft, this regulation is *also for the safety of traffic on the runways*.

 C) Wrong

In addition to the safety of traffic on runways, this regulation is *also for the safety of taxiing aircraft*.

 D) Wrong

Stop bars are *not necessary for fluent traffic on the apron*. In fact, they are established for the *safety of traffic on taxiways and on runways*.

256. A pilot taxiing on an aerodrome under all-weather operations CAT II/III is approaching a stop bar, represented by red lights at 3m intervals across the taxiway. When the lights are switched on, taxiing across the stop bar...

A)   is permitted only, when no aircraft is in sight

B)   is permitted only for IFR departures

C)   is not permitted

D)   is permitted as soon as a taxi instruction is received from the aircraft's owner

A) Wrong

It is *never allowed to cross these red lights* under all-wether conditions. The pilot has to stop before the lights.

 B) Wrong

It is *never allowed to cross these red lights* under all-wether conditions. The pilot has to stop before the lights.

 C) Correct

If the red lights at a stop bar are switched on, the pilot has to *hold position* under all circumstances. It is not allowed to pass this sign due to *safety reasons.*

 D) Wrong

It is *never allowed to cross these red lights* under all-wether conditions. The pilot has to *stop before* the lights. Besides this, the *aircraft's owner* is *not permitted* to send taxi instructions.

257. When holding in front of a stop bar at a CAT II/III holding position during all-weather operations, a pilot receives a take-off clearance from TWR. The red lights of the stop bar remain switched on. The pilot must...

A)   be very careful during line-up and take-off, however follow the TWR instruction without delay

B)   disregard the TWR instruction, hold position and wait for weather improvement, because neither take-off nor landing is permitted, as long as the stop bar is switched on

C)   inform the TWR that the light signals are not switched off and hold position until the stop bar is switched off

D)   follow the TWR instruction without arguing, because instructions via radio telephony overrule light signals

A) Wrong

It is *not allowed to pass the red lights* in this situation due to safety reasons.

 B) Wrong

Holding position is a correct action, although the pilot has to inform the tower (TWR) immediately.

 C) Correct

When *holding in front of a stop bar* at a *CAT II/III* holding position during all-weather operations and receiving a *take-off clearance* from TWR, a pilot has to *hold position* and to *inform* the tower, if the *red lights* of the stop bar remain switched on. This is regulated due to safety reasons.

 D) Wrong

In this situation, the pilot has to *hold position* and to *inform the TWR*.

258. The identification with SSR will be achieved by...

A)   transmitting for DF

B)   heading changes

C)   pressing the IDENT button

D)   switching the transponder to STBY

A) Wrong

Identification with the *VDF* (Visual Direction finder) is achieved by transmitting for *DF* (Direction Finder).

 B) Wrong

Heading changes do not facilitate identification with a transponder.

 C) Correct

Pressing the *IDENT button* on the transponder results in a special signal being transmitted *to the ground radar station*. The aircraft then usually appears in bold type on the monitor.

 D) Wrong

If the transponder is switched to *STBY* (STandBY), it does not send any signals and is only warming up (Figure 13). Therefore, *identification is not possible*.

Figure 13: Transponder and standby selector.

259. The height identification with SSR will be achieved by...

A)   squawking mode A/B

B)   transmitting for DF

C)   squawking LOW-HIGH-LOW

D)   squawking Mode A/C or S

 **A) Wrong**

Although squawking mode A is available (no altitude indication), *mode B is not available/defined*.

B) Wrong

Identification with the *VDF* (Visual Direction finder) is achieved by transmitting for *DF* (Direction Finder). Hereby only position but not altitude/heigt is indicated.

C) Wrong

It is not possible to squawk *LOW-HIGH-LOW* with a transponder.

D) Correct

The SSR is a *Secondary Surveillance Radar*, able to identify position (and altitude) of primary and secondary targets. Besides the *position*, a transponder can also encode and transmit *altitude*, if mode A/C is chosen. With *mode A* altitude information is not transmitted, whereas altitude is transmitted additionally with *mode C*. To date, additionally *mode S* will be available, enabeling the identification of each aircraft by an individual 24-bit code.

260. When switching the transponder to "STBY" ...

A)   the transponder is immediately available, if required

B)   the sensibility of the receiver is reduced

C)   the selected code is transmitting altitude information only

D)   the transponder is switched off completely

A) Correct

If the transponder is switched to *STBY* (STandBY), it does not send any signals but is *warming up* (Figure 13, page 525). It is then immediately available, if required.

 B) Wrong

It is not possible to *reduce the sensitivity of the receiver*, since the radar receiver is located on the ground.

 C) Wrong

It is not possible to *transmit altitude information only*. In *mode C* position and altitude are transmitted simultaneously.

 D) Wrong

The transponder is switched off completely with the *OFF button* (Figure 13, page 525).

 Ein Transponder sollte immer auf stand-by (STBY) geschaltet werden, um ihn im Bedarfsfall sofort nutzen zu können.

Prüfungsvorbereitung für die Privatpilotenlizenz
Band 8B: Allgemein gültiges Sprechfunkzeugnis (AZF)
2. Auflage 2009

261. If an aircraft has been cleared to land and fails to land within 5 minutes after the estimated landing time and communication cannot be established with the aircraft, ATC will...

A)   transmit an emergency message

B)   alert the search and rescue service

C)   wait for 5 minutes before taking further action

D)   assume that the aircraft is diverting to the alternate aerodrome

A) Wrong

An *emergency message* is usually transmitted from the pilot or aircraft if a direct or *severe emergency situation* exists.

 B) Correct

If an aircraft has been cleared to land and *fails to land within 5 minutes* after the estimated landing time and communication cannot be established with the aircraft, ATC will *alert the search and rescue (SAR) service.*

The SAR service covers whole Germany with aircrafts and identifies endangered aircrafts (e.g. ELT, *emergency locator transmitter*) to facilitate proper rescue missions. The SAR is coordinated in a *SAR center* 24 hours a day and 365 days a year on service.

 C) Wrong

An action is *immediately carried out* if an aircraft *fails to land* within 5 minutes after landing clearance.

 D) Wrong

This *assumption might be deleterious* and endangers both aircraft and occupants.

262. If an aircraft fails to land within 30 minutes of the estimated time of landing last reported to or estimated by air traffic services units, whichever is later, and communication cannot be established with the aircraft, ATC will...

A)   take no special action

B)   wait another 30 minutes before taking action

C)   transmit an urgency message

D)   declare the uncertainty phase

A) Wrong

The *ATC* (Air Traffic Control) will act immediately if an aircraft is missing (i.e. communication cannot be established with the aircraft).

 B) Wrong

The *ATC* (Air Traffic Control) will act immediately – and not 30 minutes later – if an aircraft is missing (i.e. communication cannot be established with the aircraft).

 C) Wrong

ATC will act *immediately* to find the aircraft. The *transmission of an urgency message* has nothing to do with the actions taken in case of a missing airplane.

 D) Correct

If an aircraft fails to *land within 30 minutes* of the estimated time of landing last *reported* to or *estimated* by air traffic services units, whichever is later, and *communication cannot be established* with the aircraft, ATC (alerting service) will declare the *uncertainty phase*.

The uncertainity phase (*INCERFA*) is a situation wherein uncertainity exists as to the safety of an aircraft or its occupants. If contanct cannot be established in the following time-course, Alert phase (*ALERFA*) and Distress phase (*DETRESFA*) are declared.

263. If an aircraft fails to arrive within 30 minutes after his ETA and ATC has no knowledge about his position, the following phase will be declared:

A)   Distress phase

B)   Alert phase

C)   Uncertainty phase

D)   Emergency phase

A) Wrong

The Distress phase (*DETRESFA*) will be declared after the uncertainity and alert phases!

 B) Wrong

The Alert phase (*ALERFA*) will be declared after the uncertainity phase.

 C) Correct

If an aircraft fails to *land within 30 minutes* of the estimated time of landing (or arriving; ETA=Estimted time of arrival) last *reported* to or *estimated* by air traffic services units, whichever is later, and *communication cannot be established* with the aircraft, ATC will declare the *uncertainty phase*.

The uncertainity phase (*INCERFA*) is a situation wherein uncertainity exists as to the safety of an aircraft or its occupants.

 D) Wrong

An *Emergency phase* is not defined.

264. Which time will be issued by ATC, if an arriving aircraft on an IFR flight has to hold over the navigation aid serving as clearance limit, when holding of more than 20 minutes is expected? The...

A)   estimated time of arrival (ETA)

B)   estimated elapsed time (EET)

C)   estimated time en-route (ETE)

D)   expected approach time (EAT)

A) Wrong

The *estimated time of arrival (ETA)* is the time at which it is estimated that the aircraft will arrive over a designated point.

 B) Wrong

The *estimated elapsed time (EET)* is the time required to proceed from one point to another (e.g. between two VORs).

 C) Wrong

The *estimated time en-route (ETE)* is the complete time required by an aircraft en-route (e.g. take off to landing).

 D) Correct

The *expected approach time (EAT)* is the time issued by the ATC at which an arriving aircraft can expect clearance *to leave the holding pattern* for further approach or descent.

265. The terminology associated with the standard holding pattern is as follows:

A)   downwind, base leg, final

B)   fix, abeam fix, outbound, holding side, inbound

C)   inbound downwind, base outbound, long final

D)   hold downwind outbound, turn inbound holding fix

A) Wrong

Downwind, base leg, and final is the terminology of a *traffic pattern*.

 B) Correct

The correct terminology of a *standard holding pattern* is as follows (Figure 14): *fix*, *abeam fix*, *outbound*, *holding side*, and *inbound*.

 C) Wrong

The terminology presented is partly from a *traffic pattern* (e.g. base outbound or long final) and therefore wrong.

 D) Wrong

The terminology presented *does not exist* in a standard holding pattern.

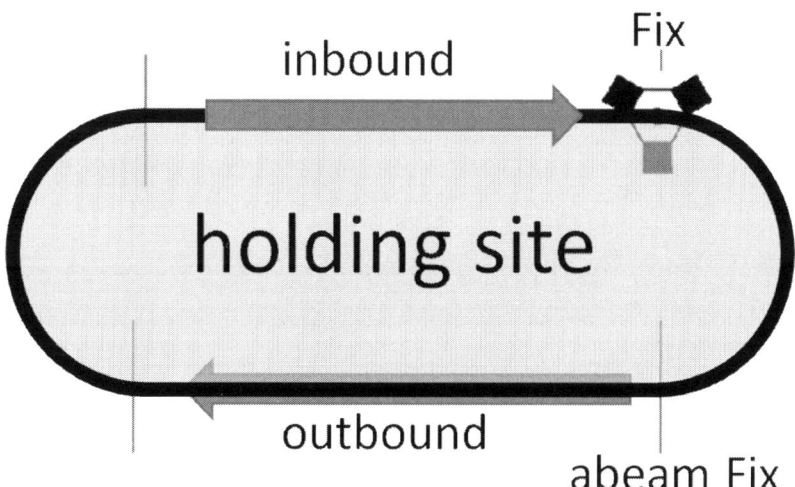

Figure 14: Terminology associated with the standard holding pattern

266. The outbound timing in a holding pattern shall begin...

A)   10 seconds before reaching the fix

B)   after having completed the turn over the fix

C)   at the convenience of the pilot

D)   over or abeam the fix, whichever is later

A) Wrong

The *outbound timing* has to begin *exactly over* or *abeam the fix*. It is not allowed to start outbound timing before overflying these points.

 B) Wrong

The "*abeam position*" is the point when having completed the turn over the fix. Therefore the answer is only partly correct, but time may also start with overflying the fix.

 C) Wrong

The *outbound timing* has to begin *exactly over* or *abeam the fix* not at the convenience of the pilot.

 D) Correct

The *outbound timing* has to begin *exactly over* or *abeam the fix* (Figure 14 or Figure 15, pages 539 or 547).

 Die Zeit zum Flug outbound beginnt exakt über oder querab des Fix.

267. What is the prescribed maximum indicated airspeed for an aircraft entering a holding pattern at FL 140 or below?

A)   240kt

B)   220kt

C)   265kt

D)   230kt

 **A) Wrong**

The presented airspeed of *240 kt* is incorrect.

B) Wrong

The presented airspeed of *220 kt* is incorrect.

C) Wrong

The presented airspeed of *265 kt* is incorrect.

D) Correct

The prescribed *maximum indicated airspeed (IAS)* for an aircraft entering a holding pattern *at FL 140 or below is 230 kt.*

 Die Maximalgeschwindigkeit in einem Holding in einer Höhe von maximal FL 140 beträgt maximal 230 kt.

268. What is the prescribed maximum indicated airspeed for an aircraft entering a holding pattern above FL 200?

A)   300 kt

B)   320 kt

C)   265 kt

D)   230 kt

A) Wrong

The presented airspeed of *300 kt* is incorrect.

 B) Wrong

The presented airspeed of 3*20 kt* is incorrect.

 C) Correct

The prescribed *maximum indicated airspeed (IAS)* for an aircraft entering a holding pattern *at FL 150 or above is 265 kt.* Therefore, airspeed in FL 200 is also *265 kt.*

 D) Wrong

The presented airspeed of *230 kt* is incorrect *in FL150 or above.* The prescribed *maximum indicated airspeed (IAS)* for an aircraft entering a holding pattern *at FL 140 or below is 230 kt.*

 Die Maximalgeschwindigkeit in einem Holding in FL 150 oder darüber beträgt 265 kt.

269. What is the outbounding time in a holding pattern up to FL 140?

A)   1 minute

B)   1 minute and 30 seconds

C)   2 minutes

D)   30 seconds

A) Correct

The time spent on the outbound leg in a holding pattern *up to FL 140* is exactly *one minute* (Figure 15). Above FL 140 the time spent on the outbound leg is longer (1:30 min).

B) Wrong

At or above FL 150 the time spent on the outbound leg is *1 minute and 30 seconds*.

C) Wrong

The time spent on the outbound leg is *never 2 minutes*.

D) Wrong

The time spent on the outbound leg is *never 30 seconds*.

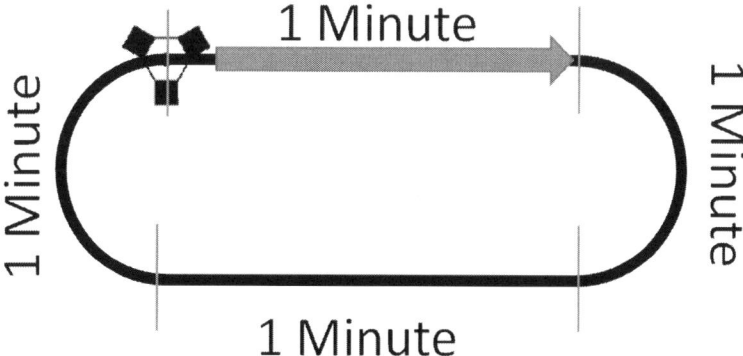

Figure 15: Outbounding time in a holding pattern up to FL 140

270. A standard holding pattern at FL 150 or above is to be flown outbound:

A)   1 minute and 30 seconds

B)   1 minute

C)   2 minutes

D)   2 minutes and 30 seconds

A) Correct

At or above FL 150 the time spent on the outbound leg is *1 minute and 30 seconds* (Figure 16).

 B) Wrong

The time spent on the outbound leg in a holding pattern *up to FL 140* is exactly *one minute.*

C) Wrong

The time spent on the outbound leg is *never 2 minutes.*

D) Wrong

The time spent on the outbound leg is *never 2:30 minutes.*

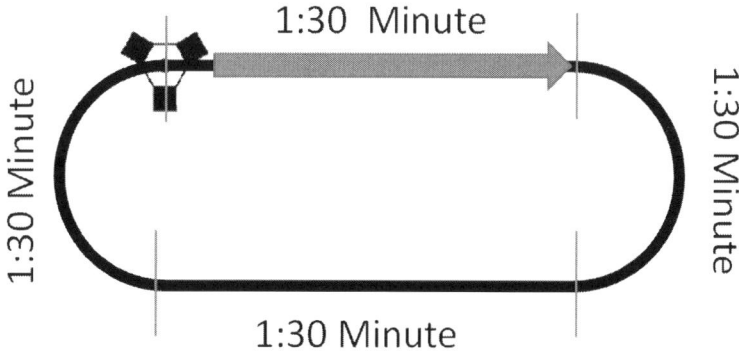

Figure 16: Outbounding time in a holding pattern at or above FL 150

271. What is the advantage of a VORTAC, compared to a VOR, if the aircraft is equipped with respective receivers?

A) The airborne equipment indicates distance and direction in respect to the position of the VORTAC

B) The TO/FROM indication can be disregarded

C) The airborne equipment indicates direction and altitude/ flight level

D) The airborne equipment indicates direction and IAS

A) Correct

The *VORTAC* (=VOR plus TACAN) is a *combination* of a *VOR* (very high frequency omnidirectional radio range) and a *TACAN* (tactical air navigation; *military station*) with the advantage of *indicating distance and direction* in respect to the position of the VORTAC. The indication is only available, if the aircraft is equipped with a *special onboard receiver*. The VORTAC (Figure 17) works comparable to a VOR/DME station.

 B) Wrong

The *TO/FROM indication* should not be disregarded even when using a VORTAC.

 C) Wrong

The *DME* but not the VORTAC indicates *direction* and *altitude/flight level*.

 D) Wrong

In contrast to the DME, the *airborne equipment* for a VORTAC does not indicate *IAS* (Indicated Air Speed).

Figure 17: VORTAC ground station (Quelle: DFS Deutsche Flugsicherung GmbH)

272. The DME receiver provides the pilot with information on...

A)   azimuth

B)   height

C)   weather

D)   distance

A) Wrong

The azimuth is not provided by a DME (Distance Measuring Equipment).

 B) Wrong

The altitude but not the height is indicatd by a DME.

 C) Wrong

Weather information is not provided by a DME.

 D) Correct

Since the DME (Distance Measuring Equipment) uses transit time from the station to the device, it allows to measure and to indicate the distance from the aircraft to the station. Additionally, airspeed and time to the station are often displayed.

273. What is the identification of the MM?

A)   Dashes

B)   Dots and dashes

C)   Continuous waves

D)   Dots

 **A) Wrong**

Solely dashes are not used as to identify a middle marker.

 **B) Correct**

The auditive sign for a middle marker (MM) consists of dots and dashes. It is derived from the Morse code for "MM".

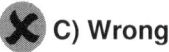 **C) Wrong**

Continuous waves are not used for the identification of any marker.

 **D) Wrong**

Solely dots are not used as to identify a middle marker.

 Der Überflug des Middle Marker (MM) wird durch lange und kurze Töne kenntlich gemacht.

274. The passage of the middle marker is indicated to the pilot by...

A)   a red light

B)   an amber (yellow) light

C)   a purple (blue) light

D)   a white light

A) Wrong

The flashing of a *red light* is usually a *warning* or *emergency* signal.

 B) Correct

The passage of the *middle marker* is indicated to the pilot by an *amber (yellow) light and an aural signal of 1500 Hz* (Table 12, page 557).

 C) Wrong

The passage of the *outer marker* is indicated to the pilot by a *purple (blue) light and an aural signal of 400 Hz* (Table 12, page 557).

 D) Wrong

A *white light and an aural signal of 3000 Hz* indicate the inner marker.

Marker	Light	Aural signal	Carrier frequency
Inner Marker	White	3000 Hz	75 MHz
Middle Marker	Amber (Yellow)	1500 Hz	75 MHz
Outer Marker	Purple (blue)	400 Hz	75 MHz

Table 12: Classification of inner, middle, and outer marker

275. Which is the carrier frequency of the outer marker?

A)   100 MHz

B)   25 kHz

C)   75 MHz

D)   50 kHz

A) Wrong

100 MHz (UHF, ultra high frequency) is not the carrier frequency of the outer marker.

 B) Wrong

25 kHz is not the carrier frequency of the outer marker.

 C) Correct

75 MHz is the carrier frequency of the outer marker (Table 12, page 557).

 D) Wrong

50 kHz is not the carrier frequency of the outer marker.

 Der Outer Marker (OM) hat die Trägerfrequenz 75 MHz.

276. The passage of the outer marker is indicated to the pilot by...

A)   an amber (yellow) light

B)   a white light and an aural signal of 3,000 Hz

C)   a purple (blue) light and an aural signal of 400 Hz

D)   a 360° needle-swing of the ADF indicator

A) Wrong

An *amber (yellow) light* indicates the passage of the inner marker, immediately before touching down.

 B) Wrong

A *white light and an aural signal of 3,000 Hz* (= 3 kHz) indicate passing the inner marker.

 C) Correct

The passage of the *outer marker* is indicated to the pilot by a *purple (blue) light and an aural signal of 400 Hz* (Table 12, page 557).

 D) Wrong

A *360° needle-swing of the ADF indicator* does not indicate the passage of a marker.

 Der Überflug des Outer Markers (OM) wird durch ein violettes Licht kenntlich gemacht.

277. The passage of a locator beacon (LO) is indicated to the pilot by...

A)   the change of the TO/FROM indicator

B)   the flashing of a purple light

C)   a 180° needle-swing of the ADF indicator

D)   the flashing of a red light and an aural signal

 **A) Wrong**

The passage of a *VOR* is indicated by the change of the *TO/FROM indicator.*

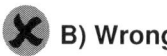 **B) Wrong**

Purple lights are not indicating locator beacons, but an indication for passing the outer marker.

 **C) Correct**

Locator beacons (LO) are *non-directional beacons* (NDB) in one of the approach segments. The passage of such a beacon is indicated to the pilot by a *180° needle-swing of the ADF indicator.*

 **D) Wrong**

The flashing of a *red light* and an *aural signal* is usually a *warning* or *emergency* signal.

 Der Übedrflug eines LOB wird durch eine Drehung der ADF-Nadel um 180° signalisiert.

278. Name all parts of a standard instrument approach procedure:

A)   Initial approach and final approach

B)   Arrival route, initial approach, intermediate approach, final approach, missed approach

C)   Final approach and missed approach

D)   Intermediate approach, final approach and missed approach

A) Wrong

In the parts listed, *arrival route, intermediate approach, and missed approach are missing.*

 B) Correct

The standard instrument approach consists of the *arrival route, initial approach, intermediate approach, final approach, and missed approach* (Figure 18).

 C) Wrong

In the parts listed, *arrival route, initial approach, and intermediate approach are missing.*

 D) Wrong

In the parts listed, *arrival route, and initial approach are missing.*

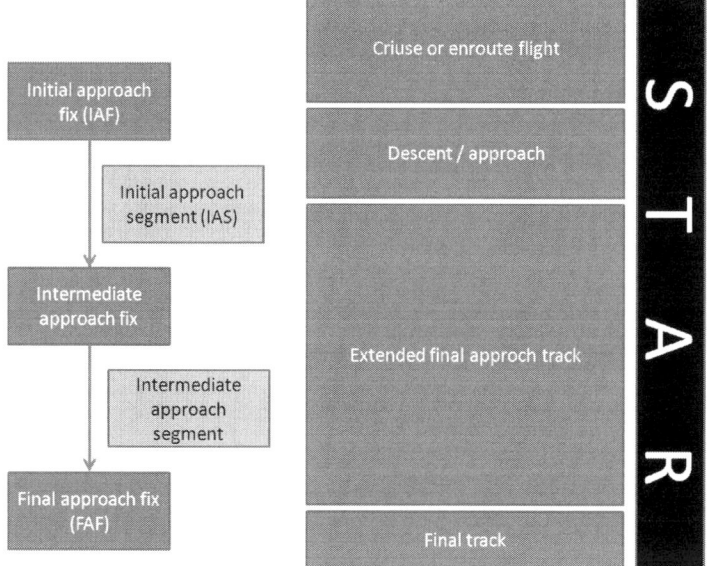

Figure 18: Parts of the standard instrument appproach

279. What are the criteria for the different aircraft categories during an instrument approach?

A)   Range of speed for initial approach

B)   Range of final approach speeds

C)   Maximum speeds for missed approach

D)   Speed at threshold based on 1.3 times stall-speed in the landing configuration at maximum certificated landing mass

A) Wrong

The *range of speed* is not a criterion for the different aircraft categories during Instrument approach.

 B) Wrong

The *range of final approach speed* is not a criterion for the different aircraft categories during Instrument approach.

 C) Wrong

Maximum speed for missed approach is not a criterion for the different aircraft categories during Instrument approach.

 D) Correct

Aircraft are categorized in four *different categories* (i.e. A to D). Category A aircraft (e.g. Airbus A380) are faster flying than category D aircraft (e.g. Piper PA28). The criteria to categorize aircraft are:

- *Speed at threshold*
- *based on 1.3 times stall-speed in the landing configuration*
- *at maximum certificated landing mass*

280. Which of the following approaches is a precision approach?

A)   ILS approach

B)   DME approach

C)   VOR approach

D)   ILS back-beam approach

A) Correct

A *precision approach* is essential when landing in *severe weather conditions*. It is therefore essential to have both the *course (horizontal) guidance* and additional *vertical* (slope) *guidance*. Every precision approach (e.g. *ILS approach*; instrument landing system=ILS) has a vertical guidance.

 B) Wrong

A *DME approach* (DME=distance measuring equipment) is a non-precision approach because it has *no vertical guidance*. It is often combined with a localizer (LLZ) approach, i.e. LLZ-DME approach.

 C) Wrong

A *VOR-approach* is a *non-precision approach* and therefore has no vertical guidance. It is often combined with a DME approach, i.e. VOR-DME approach.

 D) Wrong

An *ILS back-beam approach* is no official procedure for an approach and is thus *not a precision approach*!

 DME-, VOR- und ILS-backbeam-Approaches sind Non-precision Anflüge!

281. Which of the following approaches includes a vertical guidance?

A)   NDB approach

B)   LLZ approach

C)   ILS approach

D)   VOR approach

A) Wrong

A *NDB-approach* is a *non-precision approach* and therefore has no vertical guidance.

 B) Wrong

A *LLZ-approach* (Localizer approach) is a *non-precision approach* and therefore has no vertical guidance.

 C) Correct

A *precision approach* is essential when landing under *severe weather conditions*. It is therefore essential to have besides the *course (horizontal) guidance* and additionally *vertical* (slope) *guidance*. Every precision approach (e.g. *ILS approach*; instrument landing system=ILS) has a vertical guidance.

 D) Wrong

A *VOR-approach* is a *non-precision approach* and therefore has no vertical guidance.

 Beim NDB-, LLZ- und VOR-Anflug existiert keine vertikale Anflughilfe. Daher sind diese Nicht-Präzisionsanflüge.

282. Which of the following statements is correct?

A)   A precision approach has no vertical guidance

B)   A VOR approach has a vertical guidance

C)   A precision approach has a vertical guidance

D)   A LLZ approach has a vertical guidance

A) Wrong

A *precision approach* is essential when landing in *severe weather conditions*. Hence it is not true, that it has no vertical guidance. Only non-precision approaches lack vertical guidance (e.g. Localizer-DME Approach; LLZ-DME).

 B) Wrong

A *VOR-approach* is a *non-precision approach* and therefore has no vertical guidance.

 C) Correct

A *precision approach* is essential when landing in *severe weather conditions*. It is therefore essential to have both the *course (horizontal) guidance* and additional *vertical* (slope) *guidance*. Every precision approach (e.g. *ILS approach*; instrument landing system=ILS) has a vertical guidance.

 D) Wrong

A *localizer* (LLZ) *approach* (e.g. Localizer-DME Approach; LLZ-DME) is a *non-precision approach* and therefore has no vertical guidance.

 Beim Präzisionsanflug gibt es immer eine vertikale Anflughilfe (z.B. Glidepath).

283. Which component of the ILS provides the pilot with electronic course guidance?

A)   Localizer

B)   Glide path

C)   Marker beacon

D)   Approach lighting system

A) Correct

The *course* is the projected line on the ground flown by an aircraft. The *ILS* (Instrument Landing System) is usually used for an *instrument approach*. The *ILS-indicator* provides information on the *postion* of the aircraft compared to the required course (*localizer*, vertical line on the indicator) and the *glide path* (horizontal line on the indicator). If the vertical line is left of the center, the aircraft has to navigate more to the left and is therefore right of the required course.

 B) Wrong

The *glide path* of the ILS-Indicator provides information on the *correct altitude* in reference to the distance to the airport. If the glide path indicator is above the *horizontal center*, the aircraft is to low; if the glide path is below the horizontal center, the aircraft must descent to reach the correct altitude.

 C) Wrong

The *marker beacon* reports overflying a specified marker, i.e. point on the ground.

 D) Wrong

The *approach lighting system* is a group of visual recognizable lights on the ground in the short final approach.

284. Which instrument approach procedure segment leads an aircraft to the extended final approach track?

A)   The final approach segment

B)   The initial approach segment

C)   The intermediate approach segment

D)   The STAR

A) Wrong

The final approach segment does not lead the aircraft to the extended final approach track. Instead, the initial approach segment (IAS) leads an aircraft to the extended final approach track, which usually is in the very late phase before landing.

 B) Correct

The initial approach fix (IAF) is the point where the initial approach segment (IAS) begins. This segment leads an aircraft to the extended final approach track (EFAT).

 C) Wrong

The intermediate approach segment begins at the intermediate approach fix, which follows the initial approach segment.

 D) Wrong

The STAR (Standard Terminal Arrival Route, Standard Arrival Route; Figure 19) is a published procedure followed by an aircraft flying IFR just before arriving at a destination airport. It usually covers the phase of flight that lies between the top of descent from cruise or enroute flight and the final approach for landing on a runway.

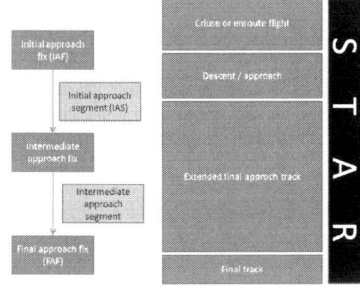

Figure 19: Standard (Terminal) Arrival Route and segments.

285. Which information gets a pilot from an air traffic controller during a SRA approach?

A)   Only headings and altitude information

B)   Radar vectors only

C)   Course corrections in regard to runway center line, distance information and altitude information

D)   Only altitudes and distance information

A) Wrong

In addition to heading and altitude information, a pilot also may request e.g. the distance to the threshold.

 B) Wrong

Besindes radar vectors, a pilot may request altitude and distance information also.

 C) Correct

During a SRA approach (surveillance radar approach) a pilot may request multiple information concerning his position and further guidance to the touch down point. This data are e.g. radar vectors in regard to the runway center line or distance information (NM) and the present altitude (ft).

 D) Wrong

Besides altitudes and distance information, a pilot may also receive radar vectors (i.e. headings to the runway).

286. The OCA refers to:

A)   Threshold

B)   MSL

C)   Field elevation

D)   QFE

A) Wrong

The OCA (*obstacle clearance altitude*) is the altitude at which no obstacles have to be expected. It is not referred to the *threshold*, since it then had to be a *height*.

 B) Correct

The *OCA* (obstacle clearance altitude) is the *altitude* at which no obstacles have to be expected. The reference for the OCA is the *mean sea level (MSL)*.

 C) Wrong

The OCA does not refer to the *field elevation*, since it then had to be a *height*.

 D) Wrong

The OCA does not refer to the *QFE* (atmospheric pressure at field elevation), since it then had to be a *height*.

 Die OCA bezieht sich immer auf MSL!

287. The pilot shall discontinue an instrument approach and initiate the missed approach procedure, if he is unable to terminate the approach to land with visual reference to the ground, when he has reached...

A)   the minimum safe height

B)   the minimum radar vectoring altitude (MRVA)

C)   the minimum sector altitude (MSA)

D)   the obstacle clearance limit (OCL)

A) Wrong

The *minimum safe height* (MSH) is the height at which no obstacles are present and hence flying is safe.

 B) Wrong

The *minimum radar vectoring altitude* (MRVA) is the minimum altitude with radar coverage and at which radar vectors are available.

 C) Wrong

The *minimum sector altitude* (MSA) is the minimum altitude at which no obstacles are present and hence flying is safe. This altitude depends on the heading being flown.

 D) Correct

A *minimum altitude* is defined for every *instrument approach*, at which the *missed approach procedure* has to be *initiated*, if the approch can not be completed with visual reference. This minimum altitude is named "*obstacle clearance limit (OCL)*".

Example: The OCL is 800 ft for an instrument approach. If the pilot is flying at 1,000 ft and has no visual reference to the ground, he may proceed down to the OCL. If the ground is visible at 800 ft, the pilot may land; if he has no visual reference at 800 ft, the missed approach procedure has to be initiated ("go around").

288. What is the Mode S Aircraft Identification?

A)   Call sign according to field 7 of the flight plan

B)   The registration number of the aircraft

C)   Flight number according to the airlibe time table

D)   The name of the aircraft operating agency

A) Correct

Mode S transponders allow the *individual identification of an aircraft* on the radar screen. Therefore, the call sign of the aircraft is transmitted. General Aviation aircrafts must be equipped with a Mode S transponder when entering airspace A, B, C, or D.

 B) Wrong

The *registration number of the aircraft* is not transmitted with a Mode S transponder.

 C) Wrong

Not only *commercial airline aircrafts* but also aircraft in General Aviation must be equipped with a Mode S transponder when entering *airspace A, B, C, or D.*

 D) Wrong

The *name of the aircraft operating agency* is not relevant for Mode S transponders.

289. What is the transition to final approach?

A)   Standard IFR-approach

B)   A non-precision approach

C)   An overlay to radar vector pattern procedure

D)   A SRE approach

A) Wrong

A transition to final approach is not a *standard IFR-approach*. A standard IFR-approach uses *navigation fixes* (e.g. initial approach fix, IAF) for maneuvering.

 B) Wrong

A transition to final approach is a *precision approach* due to radar vectors.

 C) Correct

A standard IFR-approach uses *navigation fixes* (e.g. initial approach fix, IAF) for maneuvering. The transition to final approach uses *radar vectors* for navigation instead of defined navigation fixes. It is an overlay to radar vector patterns.

 D) Wrong

A transition to final approach is not a *SRE approach*.

290. What is necessary for the execution for a transition to final approach?

A)   ILS

B)   ADF

C)   Database

D)   VOR/DME

A) Wrong

To execute a transition to final approach, a database with the overlay of a radar vector pattern procedure is necessary. Therefore, an *ILS is not necessary* for the execution of a transition to final approach.

 B) Wrong

To execute a transition to final approach, a database with the overlay of a radar vector pattern procedure is necessary. Therefore, an *ADF is not necessary* for the execution of a transition to final approach.

 C) Correct

Since the *overlay of a radar vector pattern procedure* is necessary to execute a transition to final approach, a *database* is required with the navigation aids integrated.

 D) Wrong

To execute a transition to final approach, a database with the overlay of a radar vector pattern procedure is necessary. Therefore, an *VOR or DME is not necessary* for the execution of a transition to final approach.

Figures

Tables

Index

N

O

P

An:
aeromedConsult Hinkelbein Neuhaus GbR
z.Hd. Herr Dr. J. Hinkelbein
Goethestr. 7
76771 Hördt

Sehr geehrte Leserinnen und Leser,

um eine andauerne Verbesserung dieses Buches zu erreichen, bitten
wir Sie um Ihre Mithilfe. Bitte teilen Sie uns Ihre Meinung mit, denn
nur so kann eine stetige Verbesserung erreicht werden.
Der Einfachheit halber dürfen Sie an geeigneter Stelle Schulnoten
verteilen!

Insgesamt war ich mit dem Buch zufrieden (Note 1-6)...............

Der Text war verständlich (Note 1-6)..

Die Anzahl der Abbildungen war angemessen (Note 1-6)...........

Der Preis ist angemessen (Note 1-6)..

Wie sind Sie auf das Buch aufmerksam geworden?...................

....

Was hat Ihnen besonders gut gefallen? ....

....

....

Was sollten wir in der Folgeauflage verbessern bzw. ändern? ...

....

....

Sonstige Anmerkungen?...

....

Gerne können Sie auf der Rückseite Ihre Adresse vermerken, aber
uns auch anonym zusenden – ganz wie Sie möchten.

Vielen Dank für Ihre Mühe und Unterstützung!   (8B, 2.Aufl.)

Fachbuch

Flugmedizin und Flugpsychologie für die Privatpilotenausbildung
(Paperback, DIN A5)
2. Auflage 2009

ISBN-13: 978-3-941375-05-5

Preis: 29,95 €

mit insgesamt:
325 Seiten
50 Abbildungen
10 Tabellen

Kontaktadresse und Bestellung:
aeromedConsult Hinkelbein Neuhaus GbR
Fax: +49 (0) 3212 1184690
eMail: info@aeromedconsult.de • www.aeromedconsult.de

Reihe
Prüfungsvorbereitung
für die Privatpiloten-
lizenz

aeromedConsult

Band 1: Luftrecht

Fragenkatalog: Stand 2009

2. Auflage 2010

ISBN-13: 978-3-941375-12-3

Preis: 49,95 €

mit insgesamt:
804 Seiten
13 Abbildungen
3 Tabellen

Band 2A: Navigation
(PPL-A, PPL-N)

Fragenkatalog: Stand 2009

2. Auflage 2010

ISBN-13: 978-3-941375-15-4

Preis: 49,95 €

mit insgesamt:
722 Seiten
20 Abbildungen
1 Tabelle

Reihe
Prüfungsvorbereitung
für die Privatpiloten-
lizenz

aeromedConsult

Band 2B: Navigation (PPL-C)

Fragenkatalog: Stand 2009

2. Auflage 2010

ISBN-13: 978-941375-16-1

Preis: 39,95 €

mit insgesamt:
401 Seiten
5 Abbildungen
1 Tabelle

Band 3A: Meteorologie (PPL-A, PPL-N)

Fragenkatalog: Stand 2009

1. Auflage 2010

ISBN-13: 978-3-941375-14-7

Preis: 49,95 €

mit insgesamt:
ca. 850 Seiten
Abbildungen
Tabellen

Reihe
Prüfungsvorbereitung
für die Privatpiloten-
lizenz

Band 6A: Verhalten in besonderen Fällen (PPL-A, PPL-N)
Fragenkatalog: Stand 2009

2. Auflage 2009

ISBN-13: 978-3-941375-04-8

Preis: 49,95 €

mit insgesamt:
548 Seiten
25 Abbildungen
6 Tabellen

Band 7: Menschliches Leistungsvermögen
Fragenkatalog: Stand 2009

2. Auflage 2009

ISBN-13: 978-3-941375-06-2

Preis: 29,95 €

mit insgesamt:
272 Seiten
18 Abbildungen
4 Tabellen

Reihe
Prüfungsvorbereitung
für die Privatpiloten-
lizenz

aeromedConsult

Band 8A: Beschränkt gültiges Sprechfunkzeugnis (BZF)

Jochen Hinkelbein, Christopher Neuhaus
2. Auflage - Stand 2009

Band 8A: Beschränkt gültiges Sprechfunk-zeugnis (BZF)
Fragenkatalog: Stand 2009

2. Auflage 2009

ISBN-13: 978-3-941375-07-9

Preis: 49,95 €

mit insgesamt:
536 Seiten
38 Abbildungen
13 Tabellen

Band 8B: Allgemein gültiges Sprechfunkzeugnis (AZF)

Jochen Hinkelbein, Christopher Neuhaus
2. Auflage - Stand 2009

Band 8B: Allgemein gültiges Sprechfunk-zeugnis (AZF)
Fragenkatalog: Stand 2009

2. Auflage 2009

ISBN-13: 978-3-941375-08-6

Preis: 49,95 €

mit insgesamt:
610 Seiten
20 Abbildungen
12 Tabellen

Reihe

Prüfungsvorbereitung
KOMPAKT

aeroMedConsult

(ohne Fragen! Nur Antworten, Lösungen und Kommentare!)

KOMPAKT
Band 7: Menschliches
Leistungsvermögen
Fragenkatalog: Stand 2009

1. Auflage 2010
ISBN-13: 978-3-941375-20-8

Preis: 19,95 €

mit insgesamt:
135 Seiten
18 Abbildungen
4 Tabellen

KOMPAKT
Band 8A: Beschränkt
gültiges Sprechfunk-
zeugnis (BZF)
Fragenkatalog: Stand 2009

1. Auflage 2010
ISBN-13: 978-3-941375-17-8

Preis: 29,95 €

mit insgesamt:
270 Seiten
38 Abbildungen
13 Tabellen

Reihe

Prüfungsvorbereitung KOMPAKT

aeromedConsult

(ohne Fragen! Nur Antworten, Lösungen und Kommentare!)

Prüfungsvorbereitung
für die
Privatpilotenlizenz

Band 8B: Allgemein gültiges
Sprechfunkzeugnis (AZF)

Jochen Hinkelbein, Christopher Neuhaus
1. Auflage - Stand 2009

KOMPKAT
Band 8B: Allgemein gültiges Sprechfunkzeugnis (AZF)
Fragenkatalog: Stand 2009

1. Auflage 2010
ISBN-13: 978-3-941375-18-5

Preis: 29,95 €

mit insgesamt:
304 Seiten
20 Abbildungen
12 Tabellen

Demnächst
in unserem Verlags-programm

Achten Sie auf die zukünftigen Auflagen in Anlehnung an den aktualisierten Fragenkatalog 2009!

Die komplette und preisgünstige KOMPAKT-Reihe basierend auf dem Fragenkatalog 2009

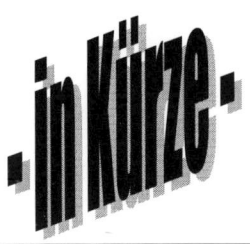

Prüfungsvorbereitung für die Privatpilotenlizenz, Band 4: Aerodynamik
(Pappeinband, DIN A5)

Prüfungsvorbereitung für die Privatpilotenlizenz, Band 5: Technik
(Pappeinband, DIN A5)